Turn Off the Fat Genes

Also by Neal Barnard, M.D.

Food for Life

Eat Right, Live Longer

Foods That Fight Pain

Turn Off the Fat Genes

The Revolutionary Guide
to Losing Weight

Neal Barnard, M.D.

Menus and Recipes by Jennifer Raymond

 THREE RIVERS PRESS • NEW YORK

Published by Three Rivers Press, New York, New York.
Member of the Crown Publishing Group.

Random House, Inc. New York, Toronto, London, Sydney, Auckland
www.randomhouse.com

THREE RIVERS PRESS and the Tugboat design are
registered trademarks of Random House, Inc.

Originally published in hardcover by Harmony Books in 2001.

Printed in the United States of America

Design by Lynne Amft

Library of Congress Cataloging-in-Publication Data
Barnard, Neal D., 1953–
Turn off the fat genes / Neal D. Barnard — 1st ed.
Includes bibliographical references and index.
1. Weight loss.   2. Obesity—Genetic aspects.   3. Leptin.   I. Title.
RM222.2.B3825   2001
616.3'98042—dc21
00-033466

ISBN 0-609-80904-0

10 9 8 7 6 5 4

First Paperback Edition

A Note to the Reader

M Y GOAL is to provide you with information on the power of foods for health. However, neither this book nor any other can take the place of individualized medical care or advice. If you have any medical condition, are overweight, or are on medication, please talk with your doctor about how dietary changes, exercise, and other medical treatments can affect your health.

The science of nutrition grows gradually as time goes on, so I encourage you to consult other sources of information, including the references listed in this volume.

With any dietary change, it is important to ensure complete nutrition. Be sure to include a source of vitamin B_{12} in your routine, which could include any common multivitamin, fortified soy milk or cereals, or a vitamin B_{12} supplement of five micrograms or more per day.

I wish you the very best of health.

Contents

Part One

The Search for the Fat and Thin Genes

Part Two

Manipulating Gene Action

Part Three

Menus and Recipes

Acknowledgments

I OWE an enormous debt of gratitude to the people who made this book possible, especially Peter Guzzardi, for his keen editorial vision and constant support; Patti Breitman, for her never-ending commitment to bringing healthful nutrition into the public eye; and Jennifer Raymond, whose menus and recipes translated nutritional theory into practice in a delightful way. Brian Belfiglio and Wendy Schuman were the essential bridges that helped the press and public understand the importance of our scientific work. Elizabeth Matthews and Kathryn Henderson kept this project on track in the many phases through which ideas turn into print. Suzy Parker provided beautiful illustrations.

Andrew Nicholson, M.D., launched our first research efforts showing the power of diet on weight control in our diabetes study conducted with Mark Sklar, M.D., of Georgetown University. Anthony Scialli, M.D., also of Georgetown, has been a source of expert objectivity and irreplaceable support as a coprincipal investigator in several of our nutrition studies. Wayne Miller, Ph.D., and Jolie Glass, M.S., of George Washington University conducted sophisticated metabolic analyses monitoring our research participants' progress, and Judy Harris gave them culinary inspiration in her innovative cooking classes. Paul Poppen, Ph.D., skillfully conducted the statistical analyses that separated significant scientific findings from chance events.

Special thanks to Dean Ornish, M.D., for his research efforts that have helped conquer seemingly insurmountable health challenges, and for inspiring readers of this book with his kind foreword. Adam Drewnowski, Ph.D., an innovative researcher uncovering the links between taste and human DNA; David Bassett, Jr., Ph.D., an expert in exercise genetics; and Claude Bouchard, Ph.D., a world leader in studies of genetic effects on obesity patiently answered my many questions about heredity's role in weight control and the extent to which it can be changed.

PCRM staffers Suzanne Bobela, Andrew Breslin, Simon Chaitowitz, Jennifer Drone, Matthew Fritts, Laurice Ghougasian, A. R. Hogan, Kathryn Kuhn, Mindy Kursban, Billy Leonard, and Miyun Park greatly facilitated the conduct of our research, while Nabila Abdulwahab; Deniz Corcoran; Sossena Dagne; Claudia Delman; Godfrey Fernando; Doug Hall; Peggy Hilden; Laurel Kadish; Jennifer Keller, R.D.; Kristine Kieswer; Amy Lanou, Ph.D.; Adaora Lathan-Sanders; Lisa Lynch; Meredith Morrisette; and Brie Turner, M.S., R.D., kept our educational efforts moving forward.

Most important of all, the participants in our research studies at the Physicians Committee for Responsible Medicine gave an enormous amount of time and energy as we investigated the power of foods for health, and it is to them this book is dedicated.

Foreword

HERE'S THE GOOD NEWS: The same diet that is best for losing weight is also optimal for enhancing your health and well-being.

Neal Barnard, M.D., has been a pioneer for many years in advocating the benefits of a low-fat, whole-foods, plant-based diet. In the studies that my colleagues and I at the nonprofit Preventive Medicine Research Institute have conducted over the past twenty-four years, we found that this type of diet may stop or reverse the progression of even severe coronary heart disease. Other studies have shown that eating this way may reduce the risk of a variety of other chronic diseases, including diabetes, hypertension, obesity, osteoarthritis, prostate cancer, breast cancer, colon cancer, and perhaps others.

However, most people don't really think anything bad will ever happen to them. They think prevention is borrrrr-ing: "I don't care if I die sooner, I want to enjoy my life."

For many people, these are choices worth making—not only to live longer, but also to live better. In the long run, fear of dying is not a good motivator, but joy of living is. Most people find that when they follow this type of diet, they feel better, have more energy, and experience an overall improved sense of well-being.

And they lose weight and keep it off.

You can lose weight on just about any diet. Keeping it off is a lot harder. A few years ago, the U.S. government reviewed all of the different weight-loss plans. They found that within a year, 66 percent of people gain back all the weight they lost, and 97 percent gain it all back within five years.

We found in our research, however, that the average person lost twenty-four pounds in the first year and had kept off more than half that weight five years later, even though they were eating more food, more frequently, than before. Without hunger or deprivation. Simply. Safely. Easily.

They not only felt better, in most cases they were better. We also found a greater reversal of heart disease after five years than after one year, and two and a half times fewer cardiac events such as heart attack, stroke, bypass surgery, and angioplasty. The more closely people followed the program, the better they were. Clearly, if you can reverse heart disease by eating this way, then you can help prevent it.

Most weight-loss plans are based on deprivation: counting calories, restricting portion sizes, and eating less food. Sooner or later, people get tired of feeling hungry, so they go off the diet, regain the weight, and usually blame themselves for not having enough discipline, willpower, or motivation, when the real problem is that they were going about it in the wrong way.

Here's a better way: If you change the *type* of food, you don't have to reduce the *amount* of food. Fat has nine calories per gram whereas protein and carbohydrates have only four calories per gram. So if you go from a high-fat diet to a low-fat, whole-foods diet, even if you eat the same amount of food, you consume far fewer calories. You feel better and you become healthier.

When you switch from a diet based on animal protein and simple carbohydrates to a whole-foods, plant-based diet, you get a quadruple benefit:

- The high fiber content of fruits, vegetables, grains, and beans reduces insulin levels, so you lose weight and lower cholesterol levels.
- When you eat less fat, you eat fewer calories without eating less food.
- You avoid the animal-based products rich in substances that cause illnesses.
- You get thousands of other plant-based substances that are protective.

Unfortunately, many people feel their problem is genetic and therefore nothing they do will help them stay slim. Weight problems may run in families—keeping appetites up and metabolisms down, and pushing effective weight loss out of reach. But genetic researchers now agree that although genes do affect our size and shape—sometimes profoundly so—we may be able to influence the effects of these genes through the diet and lifestyle choices we make each day.

In this book, Dr. Barnard describes how your unique genetic blueprint affects your tastes and appetite, builds enzymes that store fat, and sets how quickly you burn it. Although your genes build the microscopic cellular machinery that extracts fat from your bloodstream and stores it as body fat, diet changes may alter the raw materials these tiny factories use to build fat. And while other genes affect your appetite, the number of calories you may need to feel satisfied often depends on the foods you choose.

When you eat this way, you may feel better and look better. It's not all or nothing—the more you move in this direction, the more benefits you receive. And you will significantly reduce the risk of heart disease, cancer, and other illnesses rather than increasing it. You can lose weight and gain health.

Dean Ornish, M.D., Founder and President,
Preventive Medicine Research Institute
Clinical Professor of Medicine, School of Medicine,
University of California, San Francisco

Introduction

WHEN MICHELANGELO carved a sculpture, he began with a coarse tool called a point chisel, used to remove fairly large amounts of marble. As the body shape began to appear in rough form, he used a claw chisel, then smaller flat chisels to reveal a torso, arms, and legs. Finally, using rasps and files, he created the fine textures of skin and hair.

The results are timeless. With simple tools and confident hands, he brought beauty out of a formless chunk of stone.

The cells of your body are constantly sculpting your own size and shape, using tools that are more refined than any Michelangelo ever dreamed of. These tools are your genes; they are found on chromosomes in every cell of your body.

Genes are powerful. You know many people who remain slim with no effort at all. They have "thin genes." They inherited them from their parents and, in turn, will pass these traits along to their children. Other people envy them for their effortless weight control. But the fact is, we all have *thin* genes. Your body may not have put them to work effectively, but they are there. Some can block your ability to make fat, burning off calories as body heat instead. Others, like the leptin gene, are designed to tame your appetite. Still others can rev up your metabolism or help you get a better burn from exercise. These genes are waiting to be turned on.

We also have "fat genes" that encourage weight gain. They can stimulate your appetite, push your body to store fat, and interfere with your ability to burn it off. There are even genes that make you love chocolate and hate broccoli.

Everyone has both kinds of genes. Both are necessary for survival. The fat genes store energy for times when food is scarce. The thin genes shut off fat storage and let calories dissipate into the air as body heat. Some

1

genes lie dormant in early life and do not manifest themselves until older age. Some affect women more than men or vice versa.

For many of us, our cells' best sculpting tools—the most decisive genes—remain unused. While our bodies are in desperate need of an artistic touch to carve away unwanted pounds, these tools are simply gathering dust.

New research has taught us a lot about how these genes work. We have even found where some of them are located—which chromosomes they are on—although this search is by no means finished. The Human Genome Project began in 1990 as a massive effort to map all 80,000 to 100,000 genes in the human body. With a $3 billion budget from the Department of Energy and the National Institutes of Health, the project soared ahead faster than anyone expected and is now largely completed. Building on previous research, this phenomenally ambitious research project has given unprecedented glimpses into the workings of the human body. Gene maps have already revealed the exact chromosomal location of genes for cystic fibrosis, Tay-Sachs disease, muscular dystrophy, and other conditions, as well as many of the genes that influence our size and shape.

If you ignore your genes, you'll have one hand tied behind your back as you try to tackle those extra pounds. The goal of this book is to help you turn off the fat genes and turn on the thin genes. We'll do it using foods, easy exercises, and other simple strategies.

GENES ARE NOT DICTATORS

Contrary to popular opinion, scientific research reveals that genes are not dictators; they are committees. They do not give orders. They make suggestions. Genes are not rogue tyrants exerting despotic control over your waistline. Rather, they work in groups, often with subtle effects, and you can nudge them in the direction you want them to go. You can counteract the fat genes and boost your thin genes.

We think of genes as unchangeable because, when it comes to eye color or hair color, they really are decisive. If they call for blue eyes or brown hair, that's it. But the genes that establish your size and shape are much more flexible. They have to be in order to store calories in body fat and then release this energy to power your body. They need to be able to adjust your appetite and your calorie burning, depending on whether food is plentiful or not, and whether you are working hard or resting. Unlike with eye or hair color, your body has to be able to change its composition from minute to minute, and from day to day.

See your body with Michelangelo's eye. When he looked at marble, he did not simply see crystallized limestone. He saw timeless stone remnants transformed by heat and pressure into a beautiful soft mass with an inner life he could see and feel. In marble quarried from the peaks near Carrara and Pietrasanta in northern Italy, he saw pathos in Mary's face, the rage of the centaurs, David's piercing gaze, and arms, legs, and torsos emerging from rock in perfect human form.

Your body has potential that is far beyond anything the earth could pack into stone. The sculptors within your cells have an eye better than Michelangelo's and tools more precise than any chisel he ever struck with a hammer. Your capacity to shape and renew your body is not perfect, but it is far greater than you may have imagined.

Human size and form come from an inner framework of bones, overlaid with hundreds of working muscles, and a fat layer just under the skin. You already know you can change your muscles dramatically depending on how you use them. While the Y chromosome is likely to give a man greater muscle mass than women typically have, it does not turn every man into a muscle-bound giant. Depending on whether or not he exercises and what kinds of food he eats, a man can build massive muscles or he can let them turn to flab. In other words, the genes for muscle mass only make suggestions. You control how the genes are expressed.

Fat genes and thin genes work the same way. Genes for building or burning fat tissue are, for the most part, simply suggestions—tendencies—and turning them on or off depends on the choices you make.

Although our chromosomes are extraordinarily complex, for the purposes of this book you need to understand just five key gene effects:

- Taste genes determine the foods you are drawn to.
- Leptin, the appetite-taming hormone, is made by a gene on chromosome 7. It spells trouble when it's not working right.
- A gene on chromosome 8 builds LPL, the key enzyme storing fat in your cells.

- Insulin, built by a gene on chromosome 11, can stimulate an after-meal calorie burn, but it can also shut down your fat-burning in some situations. We will see how to control it.
- Genes for muscle-cell types determine whether exercise comes easy for you. If physical activity has always been a daunting prospect, a simple exercise program can change that.

Hoping for weight loss does not work. Starvation diets do not work. And hours at the gym are wasted if you do not also build the right foods into your routine. But with the right choices, you'll find a more natural way to lose weight that is much easier, too, because you'll be working in sync with your natural biology.

IS IT REALLY A GOOD IDEA TO LOSE WEIGHT?

Should we try to lose weight? Many people point out, quite rightly, that preoccupation with thinness can be destructive. When we use dangerous means to get there—drugs or starvation, for example—the cure is worse than the problem. Also, people sometimes tend to cast blame on people with weight problems, as if they are not trying their darnedest to deal with it, and that can be psychologically devastating. From my standpoint as a physician, there is no value in attaching blame to any sort of health condition. But there is every reason to look at new approaches that help us reach our goals more easily.

There are important advantages to losing weight. Being overweight shortens your life, starting when you are just 20 percent over ideal body weight. Weight kills mainly by causing diabetes, high blood pressure, and heart disease, and it also increases the risk of some forms of cancer. Even modest weight reduction helps enormously to achieve our goal of a long, healthy, active life.

MAKING IT EASY

In our research studies at the Physicians Committee for Responsible Medicine, we have helped people change their diets for various reasons. We never focus on how much anyone eats. In our diabetes study, for example, we focused entirely on *what* people ate. And we found something remarkable. People lost weight effortlessly. In twelve weeks, the average research participant lost sixteen pounds without ever counting a calorie or turning down second helpings. We were not specifically trying to promote weight loss. It just happens when you change the diet in the right way. In that same short time frame, two-thirds of our participants were able to reduce their use of medicines or stop them completely.

In a later study, we found that we could use foods to get hormones into better balance. In the process, most participants lost weight without any effort at all. After one of our participants lost about thirty pounds, she called me to ask if the weight loss would eventually quit! I reassured her that she would not keep losing weight until she blew away. Body weight reaches a new plateau—a new balance—when a change in diet causes a readjustment of basic body chemistry.

One of our more skeptical research participants found it difficult to imagine that she could lose weight without the punishing limits on calories and carbohydrates she had thought were essential to dieting. At five feet six inches, her weight had varied from 130 to 205 pounds during her adult life, and menopause had seemed to make weight control a particular challenge. She was currently about forty pounds heavier than she wanted to be. In training meetings, we recommended diet changes based on the actions of genes that store and burn fat, and she raised many questions challenging the proposed diet. However, to her surprise and delight, she lost twenty-two pounds in the first ten weeks, essentially effortlessly, and her weight loss is continuing today.

A sculptor needs time to carve a beautiful figure, and it may take you a bit of time, too, although dramatic results can occur quickly. Don't rush. Let weight come off naturally, using the simple, but powerful, tools in this book. My goal is not only to help you sculpt the body you want, but also to help you keep it that way over the long run and to enjoy the feeling of being in a healthy new body. It is not so difficult to do. Michelangelo said, "Carving a figure is simple. You have only to go down to the skin and then stop." If we let our genes sculpt a slimmer, healthier body, they will know when to stop and will let us enjoy their new creation.

I wish you the very best of success.

Part One

The Search for the Fat and Thin Genes

Chapter One

The Gene Search Pays Off

GENETICS DEALS each of us a different hand, giving us our individual strengths and weaknesses. You don't necessarily have to accept what you're given. Let me make an analogy.

Casino Square in Monaco is as beautiful a place as exists on earth. As you look up from the café veranda at the swallows soaring in the light that dapples the Mediterranean waters, the place goes to your head even before the waiter arrives with a cool drink. The old casino is as stately as ever, standing shoulder to shoulder with the Hôtel de Paris, one of the world's most luxurious addresses.

Beautiful as it is, I was oblivious to all of it. I was inside the casino, staring at a jack of clubs and a six of diamonds. The blackjack dealer had a ten up, and I was in trouble.

I grew up in Fargo, North Dakota, where low-stakes blackjack is common in restaurants and hotels. I played at a hotel whose tables benefited the local community theater and wagered as ruthlessly as I could, knowing that I might be all that stood in the way of another production of *Oklahoma!* Card-counting is not particularly difficult to learn, and a good run at the right casino can pay for a ticket to France, a rental car to Monaco, and dry cleaning for a polyester tuxedo.

This had been a good game. My bets had gradually escalated before the inevitable happened. The cards started to sour. One seemingly unbeatable hand after another crumpled as the dealer collected from a table of increasingly disgruntled players.

Sitting at "third base," I was the last to play before the dealer prepared to garrote us all. Seeing my hand and that I was considering taking another card, the other players glared and muttered in six different languages that if I was to take a hit in the vain hope of getting a four or a five, I would undoubtedly fail. Not only would I likely go over twenty-one; I might also take the critical cards that would make the dealer bust.

But with a ten up, all the dealer needed was another ten, which means any ten card or any face card at all, and his twenty would instantly wipe us all out. Drawing to a sixteen is risky, but standing was an almost certain loss.

"Carte, s'il vous plaît," I said, closing my ears to the cursing from a table full of bets ten times larger than mine.

The dealer paused to see if I was serious, then dealt a card and turned it over: a four of hearts, for twenty. I breathed a sigh of relief. The table was not appeased, however, for many gamblers believe that dangerous play casts a spell on the cards.

The dealer took his next card. A five, giving him fifteen. The players cheered in anticipation of what was to follow. He pulled a ten and busted.

The truth is, my decision was not actually the gutsy play it appeared to be. By keeping track of the cards, I knew that many small cards were left in the deck, and that the odds of my getting one were far greater than usual. The hand was untenable as it stood, and it made sense to improve it.

Some of us have healthy bodies from birth, others have bigger challenges. You cannot control the initial deal of the cards. But you don't have to settle for the hand you've got. To a degree, you can improve it. No system is perfect, and even the best players don't succeed every time. But we can take the genetic hand we have been dealt and play it to win. Working toward a healthy body is much less risky than blackjack and is far more rewarding.

THE SEARCH FOR FAT AND THIN GENES

Scientists have long searched for genes that promote thinness and others that cause weight gain. In some families, almost everyone is thin. They can eat virtually anything without it showing up on the bathroom scale. In others, just thinking about food seems to bring on the pounds. A slight frame or a stocky build in Mom and Dad is reflected in their children and sometimes their grandchildren.

It's not just size. Shape runs in families, too. If your parents carried extra weight in their hips and thighs, you are likely to find it there, too. If they carried weight around their waistlines, you may well have inherited that trait.

However, a trait that runs in a family is not necessarily genetic, hardwired forever into your chromosomes. The fact is, we give our children a lot more than DNA. We also give them recipes. We pass on our tastes for foods, ideas about the role of food in the family, and innumerable mealtime traditions. So what looks like a genetic pattern may really be habits passed from generation to generation.

The question, then, is: Are weight problems caused by our genes or by too many heavy meals? Researchers have separated the effects of genes from family habits by looking at twins. Identical twins have exactly the same genes. They started as a single fertilized egg that split in two, producing two babies. Fraternal twins, on the other hand, start out as two separate eggs. Their genes are no more alike than for any other pair of brothers or sisters.

When researchers ask sets of twins to stand on a scale, or use a tape measure to measure around their waists and hips, or check how much of their bodies are made of fat or muscle, they find that identical twins are more alike than fraternal twins. The most likely reason, of course, is that they have exactly the same genes. If they overeat, identical twins put on a similar amount of weight, and it adds to their bodies in much the same places. If they go on a diet, they lose weight similarly, much more so than fraternal twins.

Claude Bouchard and his team of researchers at Quebec's Laval University set out to deliberately overfeed a group of young men—twelve pairs of identical twins—adding an extra thousand calories to their daily diet six days a week for one hundred days. For one man, this resulted in a weight gain of only nine pounds, despite continuous overfeeding. Another, however, gained fully twenty-nine pounds. Normally, when you overeat, your body gives off heat for several hours, eliminating some of the extra calories you took in. But this man was not able to do this to any appreciable degree. The energy he took in was stored as fat, not burned as heat.[1]

The main finding of the study, however, was that each man's weight gain was fairly similar to that of his twin. If overfeeding caused a small weight change or a massive one in one man, his twin had a similar response, showing that genes play a role not only in our everyday weight, but also in how we respond to marked changes in eating habits.

Even when identical twins are separated and reared in different homes, their similarity persists. Researchers at the University of Pennsylvania and in Stockholm, Sweden, compared twins who had been separated in infancy.[2] Despite being raised apart, the identical twins had a similar body size and shape, more so than fraternal twins.

The most remarkable finding, however, was that even though the identical twins were similar, they were by no means identical in either body size or shape. Their eyes were exactly the same color, their hair color was the same, and their heights were virtually identical, but their weights were not precisely the same. Genes had an influence, but they were clearly not the last word. Even with identical genes, weight can change, sometimes dramatically.

The researchers found that it did not make any difference at all in which family the twins were raised. What determined their weight was their genes and their *current* environment, which is to say, their current eating habits. In other words, despite genetic influences, *the key factor in your weight is the type of food you are eating now.*

This is vitally important. Genes do not cut out a single predetermined body size like so many paper dolls. As we will see, your weight is affected by many different genes, and the foods you eat allow the fat genes, slim genes, or a combination to reveal their effects. Your genes permit many different body sizes, and foods let you choose which one is you at any moment in time.

THE SEARCH PAYS OFF, A LITTLE TOO WELL

The search for fat genes has not been easy or direct. Some researchers looked for genes causing weight gain in livestock or in rodents, then looked for similar genes in humans. While several human genes do resemble those that cause obesity in animals, most have turned out to have little or nothing to do with human weight problems.[3]

A more comprehensive method, called *genome scanning,* carefully examines each individual chromosome for gene patterns that can be linked to weight problems running in families.[3] In the Human Genome Project, scientists use special enzymes to slice DNA samples into tiny pieces, which are then passed through an electrical field that sorts them according to size. These pieces are reproduced and analyzed in detail using miniaturized and highly automated machinery. The end result is a detailed map of the molecules that make up each chromosome and the actions in the body to which they correspond. With these and other techniques, researchers are nailing down the genes that make us thin or heavy.

One genetic culprit was hiding on chromosome 15. A slight alteration on this chromosome causes an enormous appetite that kicks in between ages twelve and eighteen. Most teenagers are big eaters, of course, but children with this condition, called the Prader-Willi syndrome, have an appetite so far beyond normal that most end up with severe heart disease or diabetes, and few survive past age thirty.

This syndrome is rare. It occurs only once in every 25,000 people, and if you were to check earlier in an affected child's life, you would see signs that not all was well: reduced fetal movements, then sluggishness at birth, followed by slow growth and unusually small hands and feet.[1] Oddly enough, the syndrome only results when the abnormal chromosome is inherited from Dad. If it comes from Mom, the child has other problems—abnormal movements and intellectual problems—but not obesity.[1]

Another gene linked to overweight turned up on chromosome 2. This one causes not just increasing weight, but also visual and hearing problems and diabetes. Doctors call this combination the Alstrom syndrome. The gene is recessive, meaning that you have to inherit the gene from both parents to see its effects.[4]

Another fat gene was found on chromosome 3. And yet another on chromosome 4. And 5. In all, of the twenty-three pairs of human chromosomes (twenty-two autosomes, plus two X chromosomes for women and an X and a Y chromosome for men), genes linked to specific human weight problems turned up on chromosomes 2, 3, 4, 5, 7, 8, 11, 12, 15, 16, 20, and the X chromosome.[5]

A "thin gene" was identified on chromosome 7. It makes leptin, a hormone that curbs your appetite and makes your body burn calories faster. (Its name comes from the Greek *leptos,* which means thin.) Leptin is made in fat cells, and when you have enough body fat, leptin travels in your bloodstream to your brain, where it signals a slowdown in your appetite.

In 1997, English researchers reported the case of two cousins who had developed massive obesity early in life. They demanded food continuously and ate much more than their siblings. At age eight, one weighed 189 pounds, and fat tissue made up 57 percent of her body weight. She had so much trouble walking that she had to have liposuction of her legs. Her cousin was only two years old, but already weighed 64 pounds. It turned out that they shared a rare mutation blocking the leptin gene. With no leptin to curb hunger, their appetites were voracious.[6]

You are not likely to have this same gene abnormality; most of us have a working leptin signal to turn down our appetites. However, certain kinds of diets can disrupt leptin's effect. In chapter 3, we will look at ways to keep leptin working normally.

These rare abnormalities are just the tip of the iceberg of genetic influences. Most common weight problems are caused not by a single gene but by several different genes conspiring together. One might affect your appetite, another slows your metabolism, and still another influences how your body responds to exercise. Their effects may be pronounced in some people and more subtle in others. In some cases, scientists have spotted a gene's effects, but have not yet pinpointed its exact location—that is, which chromosome it is on. Nonetheless, we know what these genes are doing and we can use them wherever they may be.

FAT GENES, THIN GENES, APPLE AND PEAR GENES

Thin parents tend to have thin children. This is true even when their children are adopted into another family and raised in a different environment. A study in Denmark showed that, as adopted children reach

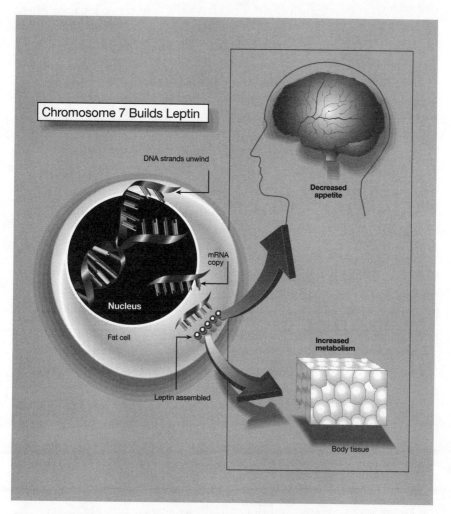

Chromosome 7 Builds Leptin

DNA strands unwind

Decreased appetite

mRNA copy

Nucleus

Fat cell

Increased metabolism

Leptin assembled

Body tissue

Just as words are strings of letters, genes are strings of molecules whose sequence communicates a message. Your cells read these codes and follow their instructions. The gene for making leptin, for example, consists of a long series of molecules linked together on chromosome 7. The cell copies this sequence onto a template, called messenger RNA, which acts as a pattern for producing leptin. Once in the bloodstream, leptin reaches the brain, where it turns down appetite, and also travels to body tissues, where it increases metabolism.

The Gene Search Pays Off

adulthood, their weight becomes less and less like that of their adoptive parents and more like that of their biological parents.[7]

In 1992, researchers in Bethesda, Maryland, found two separate genes that influence general body weight. One is active in the childhood or teen years; the other turns on after age twenty.[8] They boost your appetite, slow your calorie burning, and encourage fat storage around the internal organs.[9,10]

A gene for body *shape* was tracked down by research teams at Laval University in Quebec and at the University of Utah. It is recessive, and when you get the gene from both your parents, you'll tend to gain weight around your waistline—the so-called apple shape. Without the gene, any weight you gain tends to appear on your hips and thighs (the pear shape).[1,11]

Various gene effects have been spotted in different ethnic groups. Overall, there are many more similarities between groups than differences. For example, researchers found certain recessive genes that contribute to weight problems in both Caucasian and African-American families.[12] A similar, as yet unnamed, gene turned up in Mexican-Americans. It increases appetite and reduces calorie-burning. When inherited from both parents, it doubles body fat.[3,5]

Among the Pima Indians of Arizona, a group with common and serious weight problems, genome scanning pinpointed the Pimas' fat genes to chromosome 11 and a possible second site on chromosome 3.[13]

However, the Pimas' experience illustrates a critical point: we all have fat genes of various kinds, but they do not necessarily insist on weight gain. In fact, they can lie as dormant as seeds on dry soil until certain kinds of foods come along. The Pimas were fairly slim before the arrival of white people in the western United States. The introduction of new foods into the Pimas' diet suddenly and dramatically awakened genes that store fat. The problem foods were not necessarily those we think of as "junk food." For Pimas, even a fairly typical American menu can spell severe weight gain. The Pimas of Arizona have cousins in Mexico who have continued with a more or less traditional diet. And they do not have the obesity problems seen north of the border.

Many different genes have the potential to promote slimness or overweight. Their *expression,* in most cases, depends on the environment, and the most important part of your "environment" is inside your refrigerator, in your kitchen cupboards, and on your dinner table.

MAKING SENSE OF GENES

Let us return to our blackjack analogy for a moment. A typical game might use not merely a single deck, but four, six, or even eight decks shuffled together. If you were to try to remember every card played, the task would be impossible. Luckily for the player, counting cards is much easier. It simply means looking at the cards being dealt and keeping a simple score. For every small card on the table, you give yourself a point. For every large card, you take a point away. A higher tally indicates that plenty of small cards have been played; the big ones are still in the deck ready to give the player a more competitive hand. The endless variations on this theme are all based on the higher the count, the more likely you are to win, and you can adjust your bet accordingly.

In 1962, Edward Thorp published *Beat the Dealer*, which described card-counting in detail. It threw Las Vegas casino managers into a panic. They envisioned busloads of retirees from San Bernardino taking home bags of hundred-dollar bills. And sure enough, the blackjack tables filled up. But the casinos fought back. They had something more powerful than card-counting. They had free drinks. Once the cocktail server had laid down a second or third screwdriver, players couldn't count the fingers on their hands, let alone the cards. And there were other distractions. The slot machines clanged constantly, the craps players screamed and shouted, and the casinos removed all the clocks to help players miss their bedtimes. To win at blackjack, you had to cut through the distractions. You needed a tall glass of water and good concentration.

Similarly, weight-control genes are dizzyingly complicated. In a way, the search has paid off a little too well. Researchers now have fat- or thin-gene suspects under scrutiny on every single human chromosome, except the Y chromosome.

But for all the complexity of genes themselves, understanding gene actions is not so difficult. While there are many genes, there are only so many functions they can control. They can influence your tastes and appetite. They can affect your tendency to store fat. They can adjust your calorie-burning ability and how well exercise works for you. That's about it. Just as Carnegie Hall has hundreds of different switches controlling more than 5,000 lightbulbs, all they really do is make the room lighter or darker.

In this book, we will focus on five key gene effects. First, taste genes influence which foods you prefer. Second, the leptin system is designed to increase or decrease your appetite. Third, the fat-builder enzyme, known as LPL, plays a key role in whether calories are stored as fat or lost as body heat. Fourth, insulin, the chromosome 11 hormone, regulates your calorie-burning speed in several ways. Finally, your ability to exercise is influenced by genes for your muscle type.

Let's look briefly at each of these. The next few chapters will show you how to identify your own genetic tendencies and how foods can accentuate or counteract them.

TASTE GENES

Can you inherit a sweet tooth? No question about it. About one-fifth of your taste preference comes straight from Mom and Dad.[14] The rest depends mainly on whether you've actually tasted dark chocolate truffles, jalapeño peppers, tofu, or fried onion rings.

In taste experiments, scientists use a test substance called PROP (6-n-propylthiouracil). It is bitter, and three out of four Europeans will gladly spit it out. But one in four cannot detect it. Among Asians, few people—only about one in ten—cannot taste PROP, and this trait is similarly rare among Africans.[15] If you cannot taste it, you're not missing anything. No one ever runs to the refrigerator for second helping of PROP. The point is this: PROP is a gene marker. If your taste buds detect it, they are more sensitive to many things. Sugar will seem sweeter to you than it does to other people. Even artificial sweeteners, such as saccharin, will jump out at you more.[15]

Some people are very sensitive to PROP and can detect even a hint of bitterness in foods. So grapefruit juice and cruciferous vegetables, such as cabbage or cauliflower, leave them stone cold. "I hate vegetables," you'll hear them say. More important, PROP-tasters have an enhanced experience of sweet and fatty foods, often finding these tastes too strong. Their friends who cannot taste PROP appreciate a gooey chocolate dessert, mainly because they cannot fully taste how sickeningly sweet it is!

Being born with an incurable sweet tooth or an exaggerated distaste for sweets or any other food is not necessarily a problem. It only becomes one if your tastes lead you to the doughnut shop too many days in a row. Likewise, if a taste for fat keeps the chicken wings in your diet, you're likely to have a continuing weight challenge.

You can change your tastes. Often, they change on their own, as many people have found, including me. My father has a real sweet tooth, and I certainly inherited it. Coming home from school as a child, I often filled a bowl with chocolate ice cream. Whenever my mother made chocolate chip cookies, I was always ready for seconds and thirds. When bread came out of the toaster, I loaded it with as much jam as it could hold. But as I reached adulthood, everything changed. My sweet tooth turned off. It just quit. While sweets are still more or less tasty, they no longer seem to call out as loudly as they once did. As we'll see in chapter 3, you can change your own tastes or counter their effects, using simple techniques.

BIG APPETITES AND LITTLE ONES

Appetite is not just about taste. It is also a question of quantity. Some of us are skimpy eaters. Others are ravenous at mealtimes, barely giving our taste buds a chance to savor a bite of food before the next one is on its way.

You can see the differences. Researchers have actually timed how quickly people eat. Most people start in at a good clip but slow down as they fill up. Others have a different pattern, eating quickly and keeping this pace up from the start of the meal to the end, barely slowing down until the plate is empty.[14] This is a risky pattern, one that is linked to weight problems later on. The key, however, may not be to try to change the pattern, but to change the meal itself, as we will see.

These eating traits are not accidents. They are controlled by genes working through hormone signals. If you gain weight, for example, your fat cells produce more leptin, which reduces your appetite. If you lose weight, leptin's signal weakens, and hunger returns.[16]

The leptin gene works very differently from one person to the next. Women typically have more leptin than men, even if they have a similar amount of body fat. When researchers tested Pima Indians, they found that those with less leptin tended to gain weight, while those with more leptin tended to remain slimmer over the years.[17]

As we'll see, the amount of leptin in your blood is dramatically affected not just by genes but by the diet you follow. So, if your appetite is sometimes a bit too hearty, that does not mean that you're stuck.

THE FAT-BUILDER/FAT-BURNER GENES

As food hits your stomach, genes determine its fate. They send proteins to repair worn body parts, and carbohydrates to recharge your cells' energy. If you have eaten too much, they can turn these calories into fat or burn off the energy as heat. Certain proteins, called uncoupling proteins, disconnect your calorie storage machinery and let energy burn off as body heat, like gas burning atop an oil well.

Studies of French Canadians and Pima Indians showed that genes play a major role in controlling these proteins, and that appears to be true for everyone else as well.[18] About one-third of your ability to shed calories as body heat comes straight from your parents.[19]

Various foods differ in their ability to turn calories into heat rather than body fat. Chicken fat, for example, favors fat storage and generates almost no body heat at all, while navy beans and lentils are precisely the opposite.

Chromosome 8 holds the gene for a key enzyme, called LPL, or *lipoprotein lipase.* This enzyme waits along the tiny blood vessels that course through your body fat, and its job is to extract fat that passes by on particles in your blood, so they can pass into your fat cells for storage. Alternatively, LPL can pass fat into muscle cells to be burned as energy. As we will see in chapter 4, LPL is critical to body weight. Just as in taste and appetite, you can change a tendency to store fat into a tendency to burn away calories by the foods you select. In chapter 5, I will show you how to choose foods for a maximal burn.

CHROMOSOME 11 AND INSULIN

When I was in college, I bought an Oldsmobile that was as big as a barge. It cost $125, and it got terrible mileage. Turning the engine on, you could almost see dollar bills disappearing out of the exhaust pipe. Even before I put it into gear and drove off, it burned gas like crazy just sitting and idling.

The human body is a little bit like my old Oldsmobile. About 60 to 75 percent of all the calories you use up in a day are burned simply at rest, without your lifting a finger. Some of us have an especially fast metabolism—a fast idle, if you will—burning calories quickly, making us unlikely to gain weight. On the other hand, if you burn fuel slowly, you will be stuck with extra calories when you eat your next meal.

A slow metabolism is one of the things that made the Pima Indians vulnerable to obesity.[10,20] Presumably, their bodies' ability to hold on to body fat would stand them in good stead during times when food was scarce, and a tendency to easily store fat was no problem when their diet consisted of low-calorie foods—corn, beans, and root vegetables. But as soon as the first cheese shipments arrived, a slow metabolism simply could not keep up with the incoming load of fat. If your body burns calories slowly, it does not take many calories to gain weight.

Overall, genes control 30 to 40 percent of the speed at which your body burns calories.[19] In Quebec, Canada, researchers found that about one in every fourteen people has a gene for an especially quick metabolism that makes them resistant to weight gain almost no matter what they eat.[10] Other gene effects are less dramatic, pushing the metabolism up or down.

But whatever genes your parents gave you, you can alter your metabolism. If you have ever been on a very low calorie diet, you have seen how easy it is to push your metabolism in the wrong direction. A low-calorie diet slows your metabolism down, making weight loss more and more difficult and causing an all-too-rapid return of weight once the diet is over. You might have started out with the calorie-burning speed of a hummingbird, but a severe diet—say 500 calories a day—slows it down almost immediately.

The hormone *insulin,* coded on chromosome 11, is part of your body's system for turning on your metabolism after meals. Depending on the type of foods you choose, you can influence insulin's ability to spark a pronounced after-meal burn that releases calories as body heat, rather than storing them as fat.

The Gene Search Pays Off

EXERCISE GENES

Think about your family for a moment: Are they restless and always on the go, or are they couch potatoes? Believe it or not, when it comes to physical activity, you inherited some of your tendencies from your parents. If they were comfortable sitting still for hours, there is a good chance you are, too. About one-third of your tendency to stay active or to remain sedentary is inherited.[19]

Exercise aptitude is largely biological, too. People who love to get up at dawn and go for a five-mile run did not get that way as a result of their attitude. Their muscles are different from other people's. Scientists have taken muscle biopsies and found that people who jump into endurance sports have different muscle cells from people who would rather sit and read a newspaper. It's not sloth that makes you pull up a seat to rest, it's biology.

Having said that, a less-than-vigorous exercise aptitude can be easily tuned up by getting into a groove of more physical activity, as we will see in chapter 6. And believe it or not, even your muscle-cell type can gradually be adjusted to look more and more like those warriors whose assets came genetically.

TURNING YOUR OWN GENES ON AND OFF

Gaining weight is a remarkably simple process, which is why so many people are doing it. The genetic influences are simple, too: some push you to take in more food, others influence whether it is stored as body fat or released as heat, and still others set how well you'll burn calories at rest or during movement. Now it's time to learn how these gene groups turn on and off. In chapters 2 and 3, you'll find out your taste type and see how to influence your appetite. In chapter 4, we will look at how to block a tendency to store fat. Then, in chapters 5 and 6, we will see how to get the best possible burn and see how exercise fits in.

Chapter Two

Taste Genes: Broccoli and Chocolate

I DO NOT like broccoli," President George Bush declared on March 22, 1992. "I'm president of the United States, and I'm not going to eat any more broccoli."

Those words did not quite stir the soul like John Kennedy's "Ask not what your country can do for you," or FDR's "A day that will live in infamy," but nonetheless they were oddly memorable, perhaps because of the prompt retaliation by angry farmers who dumped forklifts full of broccoli at the White House.

Many people love broccoli. It is a favorite in North America and Europe and has an ever-growing list of healthy attributes from intrinsic cancer fighters to highly absorbable calcium. Yet some people just cannot get their taste buds to go for it. As children, they resisted their parents coaxing to "just give it a try," and as adults they shy away from the tray of crudités.

Well, you may have suspected it, and it is true: taste is genetic. Some genes favor a sweet tooth, and some genes really do make you dislike broccoli. No amount of parental arm-twisting makes those genes go away.

This should be no great surprise. We already had hints that genes influence our tastes, based on the effect of the X and Y chromosomes. Women, with two X chromosomes, are often drawn to sweets—cookies, cakes, ice cream, and chocolate—and cravings are accentuated as hormones shift during the menstrual cycle or pregnancy. Men, with an X and a Y chromosome, can have a sweet tooth, too, but more often prefer fatty, salty tastes—steak, burgers, french fries, and pizza.[1]

In this chapter, we will take a look at the vulnerabilities that go along with different taste types. If you like, you can even change your tastes, to a degree. If you are especially drawn to sweets, or to fatty or salty snacks, we'll look at how to bring these tastes back into bounds.

TASTE GENES

If genes are the microscopic artists that sculpt your body's size and shape, your genes have their own "tastes," if you will, just as any artist does, and they influence your taste in foods. Some of us have "Giacometti genes." When you look at a Giacometti statue, you want to buy it a good square meal. Alberto Giacometti used spare materials, and his figures are absurdly thin, frail, and vulnerable.

Others of us have "Henry Moore genes." Moore was keen on substance, with tastes as different from Giacometti's as English beef is from vermicelli. The arms, breast, and ample hips in his 1938 *Large Reclining Figure* are solid bronze spanning thirty feet.

If your genes prefer Moore to Giacometti—if they lead your own tastes toward fattening foods—this is not necessarily the body size and shape they will sculpt. You can teach your body a thing or two about taste. Let's take a look at the genetics beneath your taste buds.

At first, we are all pretty much the same. Before you were born, genetic blueprints set your taste buds to prefer just one taste: sweets. Their gustatory galleries cater to Norman Rockwell. Newborns are drawn to the mild sweetness of mother's milk and, later, to the fruits that are among the first solid foods. This is innate and has nothing to do with learning or bonding to the breast.

Babies are repelled by any hint of bitterness. This taste-bud programming automatically wrinkles their noses at an approaching spoonful of medicine. It will not accept broccoli, cabbage, or grapefruit for another two decades. As any parent soon learns, pushing children to taste the foods Mom and Dad like is a pointless exercise. Some adults can get excited about broccoli, mutual funds, landscaping, or deck sealant, but a child just cannot. Their taste buds are on a genetic time clock, and for now, sweet fruits are in, bitter vegetables are out.

Tastes vary enormously from one person to the next, and genetics are a big part of the reason. As we saw in the first chapter, scientists can check your ability to taste bitter substances with a test compound called 6-n-propylthiouracil, or PROP for short. Some people—about one in four in the United States—cannot taste it. About half can detect it, and another one-fourth are extrasensitive to it, what scientists call supertasters.[2] People of African or Asian heritage are more likely than Caucasians to be able to taste PROP.

If neither parent gave you the PROP taster gene, you are a "nontaster." This means you will not be bothered by bitter tastes. Grapefruits or broccoli will taste perfectly fine. But if you got the gene from one or both parents, you will be more alert to bitter tastes. Getting the gene from both parents is believed to be what makes you a supertaster. Microscopically, the surface of your tongue actually looks different from that of other people, with more tiny mushroom-shaped (what scientists call fungiform) taste papillae among your taste buds.*

IF YOU HAVE THE PROP-TASTER GENE

From the standpoint of weight control, being a PROP taster has advantages, although also some quirks. If you are a supertaster, you will detect a bitter taste in caffeine that is too subtle for others to notice. You will find Starbucks coffee unbearable unless heavily diluted with coffee creamer and sugar. Other tastes might stand out more, too. You may stop your moviegoing friends from adding extra salt to popcorn because, to you, it is salty

---

*If you are in an adventurous mood, you can actually check this at home. Using McCormick household blue food coloring, paint the front of your tongue with a cotton swab. Then, using a quarter-inch (65mm) office paper punch, cut a hole in a notecard and place it over the tip of your tongue, adjacent to the midline. You'll see your taste papillae standing out like red islands in a sea of blue. Count them. If you count fewer than twenty-three, you are likely a nontaster. If you have twenty-three to twenty-five, you are a borderline case. If you have more than twenty-five, you are likely a taster or supertaster.[3]

UNDERSTANDING YOUR TASTE TYPE

Let's take a look at your taste type. Remember, genes are not dictators, and if you were born with tastes that have led you down chocolate lane a few thousand times, you can change them, at least to a degree. In fact, you have probably already done so many times—just think of foods that left you a bit cold at first but then became more attractive as you got to know them.

Ask yourself these questions:

- Do you really dislike cabbage or brussels sprouts?
- Do you find grapefruit juice intolerably bitter?
- Do you find coffee undrinkable without coffee creamer and/or sugar?
- Do you often find some sweets are just too sickeningly sweet for your tastes?

If you answered yes to most of these questions, you are likely a PROP taster. Congratulations—your taste buds are guiding you in the right direction. If you answered no to most of these questions, you may well be a PROP nontaster, although it is also possible that you are a taster who has learned to flavor up bitter foods or has adapted to them in some other way. In the section "If You Have the PROP-Taster Gene," beginning on page 25, we look at what your taste type means to you.

enough already. You may find sweets sweeter—perhaps sickeningly so—and hot peppers hotter, if not downright painful.[3]

If you were to taste, say, green tea, you would wonder why anyone ever drank it more than once. You will not be particularly keen on unflavored soy milk because some of the natural anticancer soy compounds have a slightly bitter taste. But you will love vanilla-flavored soy milk, because you can easily detect its faint sweetness.[5]

PROP tasters can also readily pick out fats and oils in foods. Researchers at Rutgers University combined Good Seasons Italian dressing mix with oil, apple-cider vinegar, and water, making one batch that was about 10 percent oil and another about 40 percent oil. Nontasters could not tell the two apart, while the PROP tasters had no trouble doing so.[3] Luckily for them, they found the lighter versions more palatable.[6]

Turn Off the Fat Genes

PROP tasters may be able to conquer certain attractions a bit more easily than others can. Cheddar cheese will not only stack up as a fatty and calorie-dense food, it will also taste bitter.[7] And because alcohol will seem a bit bitter, too, you are less likely to develop a drinking problem.[4]

In a nutshell, sensitive taste buds make you more alert to many different tastes, particularly those that are not so healthful. Oddest of all is your reaction to sweets. In an experiment at Canada's McMaster University, researchers asked students whether they liked the taste of various sugar-water samples, some of which were weak while others were loaded with sugar. And just to be scientific about it, they did not take yes or no for an answer. They videotaped the reactions to the overly sweet drinks, which they dutifully recorded as "frowning, tongue-thrusting, whole-face grimacing, rolling of the eyes, nostrils flaring, head rearing back, and a downturned or open mouth."[8]

Slightly more than half the students disliked the especially sugary drinks. Nearly every one of these sugar haters turned out to have the PROP-taster gene. If you like sweets, you may or may not be a PROP taster, but if you find sweets overpowering, it is a good bet that the PROP-taster gene has made you extrasensitive to them. Apparently, people who like sweets do not detect the sugary flavor quite so strongly, and other flavors are able to come through. To them, chocolate tastes like chocolate, and bits of mint, fruit flavor, or tartness bring character to candies. These flavors come through for candy connoisseurs, but seem to be drowned in sickening sweetness for PROP tasters.*[8]

Although you may avoid sweets, there is one disadvantage to being a PROP taster: you are at risk for turning away from healthy vegetables and toward unhealthful foods. George Bush never ate broccoli and was keen on pork rinds and other foods not found on elegant menus. If a distaste for vegetables leads you to fatty foods, you could easily risk overweight and

---

*Apparently, two different genes determine your sensitivity to bitter foods. One determines your taste acuity in general, while the other controls bitter tastes specifically. Surprisingly, the genes for bitter tastes are also linked to other biological traits. Girls who can taste PROP reach puberty earlier than nontasters by about four months on average. The reason is unknown, but is presumed to relate to fundamental differences in hormone function.[4]

heart disease and will miss out on the vitamins and other nutrients in vegetables.

Your solution is twofold. First, choose your vegetables carefully. Many healthful vegetables will not trigger your bitter sensors. Carrots, sweet potatoes, green beans, and any of the lettuce varieties will go over perfectly well. Second, you can even learn to love spinach or cruciferous vegetables if they are prepared in the right way. Be sure to allow adequate cooking time, and try adding a splash of lemon juice. It is a mystery why adding a sour taste, such as lemon juice, eliminates the bitterness of vegetables, but it works. Fat-free salad dressings are also a good choice.

PROP tasters not only have more sensitive taste buds, but Roland Fischer, a researcher at Ohio State University, found that they are more sensitive in many other ways, too. They react to medicines at lower doses and have a lower tolerance for pain. Nontasters, on the other hand, generally needed higher doses of medication for a therapeutic effect and needed more stimulation in general to feel comfortable. They tended to be more extroverted and stimulus-seeking, while tasters were less so. He also found that couples and groups of friends are often similar in their taste patterns. Presumably, people tend to seek out friends and partners with similar personalities, which are reflected in surprisingly similar taste acuity when put to the test.[9]

IF YOU ARE GENETICALLY A NONTASTER

Some have called the inability to taste bitter substances, such as PROP, "taste insensitivity" or even "taste blindness." On the positive side, you will not run screaming if a stalk of broccoli graces your plate. Your ability to tolerate straight black coffee will amaze your friends. Your main concern is that, if your tastes are somewhat insensitive, you may have trouble noticing when things go too far. After others have pushed away the birthday cake, you might find yourself wanting another slice. You are more likely to eat high-fat dairy products, such as ice cream. Cheese tastes fine, alcohol tastes fine—their natural bitterness simply passes you by.

It may not surprise you that Yale University researchers found that PROP nontasters are more likely to gain weight. When they stand on a scale or have their waistlines or hips measured, they have accumulated more body fat, compared to tasters.[10]

Fear not. Overall, the difference in weight between tasters and nontasters is not major. A better way to think of it is that both tasters and nontasters have their own vulnerabilities, and being aware of them can be of enormous help. Tasters have to choose healthful foods that do not set off their taste alarms, and nontasters need to be sure not to overdo it in their quest to enjoy foods.

Tastes are not set in stone, and taste genes can be counteracted by experience. Whatever your type, the general nutritional guidelines in chapter 9 will suit you to a tee.

WHY DO WE HAVE TASTE GENES?

What's the point of having acute tastes? Almost certainly, it is nature's way of detecting poisons. Food that has spoiled often has a bitter or sour taste. If you can spot it before you swallow your first bite, you're better off.

Broccoli and other cruciferous vegetables (the group that includes cauliflower, cabbage, and brussels sprouts) have tiny amounts of natural toxins that repel insects or other attackers. This is not a reason to avoid them; these natural defenses are apparently neutralized by sufficient cooking. Moreover, these foods are cancer fighters and have many other health benefits. Nonetheless, the bitter taste is a warning about the natural chemical defenses of the raw plant, weak though they may be.

The bitterness of the grapefruit is an accident. Grapefruit were developed in the eighteenth century in the West Indies as a cross between an orange and a pomelo, a large citrus fruit. The breeding built into the fruit a bitter compound called naringin, which has a peculiar property: it slows your liver's ability to eliminate chemicals from your bloodstream. For example, if you drink a glass of grapefruit juice at breakfast, your liver has trouble eliminating caffeine. As a result, the buzz from your morning coffee lasts longer.* Naringin also slows down your liver's removal of some

---

*If you have the jitters after your morning coffee, the blame could go to the grapefruit juice that preceded it, which reduced your ability to eliminate caffeine. Believe it or not, a morning cigarette has exactly the opposite effect. It makes your body eliminate caffeine faster. When smokers quit, their caffeine-eliminating machinery slows to normal speed, and the amount of caffeine in their blood doubles. In fact, some of the jitters they think are related to nicotine withdrawal are actually caused by the extra caffeine jolt.

drugs, such as calcium channel blockers.[11] PROP tasters find the bitterness of grapefruit hard to take. Of course, your taste buds are not looking to avoid grapefruit. Rather, they are on the lookout for natural toxins or spoilage.

SWEETS AND SAVORIES

The PROP taste is only the beginning of our taste types. Let us go a step further and look at other specific vulnerabilities:

- Is a life without chocolate just not worth living? Do sweets call to you even when you are not the least bit hungry?
- Do savories call to you more than sweets? Is a steak or fried chicken, or a salty snack, such as potato chips, french fries, or greasy popcorn, more your thing?

When it comes to snacks, some of us lean toward sweets, others toward savories. Some of us find equal attractions in both, while others rarely snack at all. In the sections below, we will understand where these tastes might lead us. If you would like to actually change your tastes, chapter 7 includes a three-week diet makeover that does exactly that.

THE SWEET TOOTH

All children are attracted to sweets. The sweeter the taste, the better they like it. As we grow older, genetic programming turns this taste down a bit, and most adults like sugar only up to a point. In fact, taste researchers have gone to great lengths to nail down exactly what that perfect point is, and they seem to have found it: in beverages, it is five teaspoons of sugar in a cup of water, adjusting the amount a bit if other ingredients are added.[1] In solid foods, it is found with roughly even amounts of sugars and fats.

While most of us lose some of our fondness for sweets as we reach adulthood, some people have a remarkably persistent sweet tooth. If that includes you, you'll be happy to know that researchers have found that, in and of itself, a taste for sweets does not cause weight problems. If you like to sweeten up your lemonade or have a dish of canned peaches in syrup, these are not linked to weight problems.[1]

However, if you really put away a lot of sweets, you will end up with

Turn Off the Fat Genes

extra calories, about fifteen in every teaspoon of sugar. Where this becomes a real problem is when sugar acts as a Trojan horse for fat. Sugar lures you to chocolate, ice cream, cake, and pudding, and their not-so-hidden fats hold the calories that easily add to your waistline. These fats have more than twice the calories of sugar, ounce for ounce.

Trojan horses are everywhere. Take a Snickers bar, for example. Of its 300 or so calories, about half come from sugar, the part we feel guilty about. But the other half come from cocoa butter, butter, milk fat, and hydrogenated oil. Similarly, while sugar contributes 130 calories to a cup of vanilla ice cream, fat adds 130 more. A package of Hostess cupcakes holds 136 calories worth of sugar, and another 100 come from fat. As we will see in chapter 4, fat calories present a particular problem in that they are easily stored as body fat.

If you were going to blame your sweet tooth on weak will or the Chocolate Manufacturers Association, the truth is, you can blame your distant ancestors. Your taste buds are actually programmed to look for calories. When a baby tastes some new fatty food, his or her brain associates its taste and feel with dense calories and quickly adds it to a growing list of favorites.[12] An apple may be sweet, but its 74 calories from natural fruit sugars are no match for chocolate, which is dense in both sugar and calorie-rich cocoa butter. A glass of orange juice has 111 sweet calories, but a modest ice cream serving packs in well over double that number from sugar and milk fat. Taste scientists have discovered that taste nirvana occurs not from sugar alone—after all, few people would ever take a spoon to a box of sugar—but from the combination of sugar and fat.[1] The combination responds to the most basic human taste sensation and hammers it home with enough fat calories to register in the brain.

Of course, it's all well and good to know why we love chocolate so much. But what do we do about it? Before tackling that question, I need to give you one more bit of news. That is, if you thought you were a chocolate addict, you may be exactly right.

THE TRUTH OF CHOCOLATE ADDICTION
Chocolate is the single most craved food, and for understandable reasons.[16] As far as your brain in concerned, it is a drugstore hidden in a food. Not only

does it contain caffeine and a related compound, theobromine, it also has an amphetamine-like ingredient, phenylethylamine (PEA), and works like an opiate.* It turns out to be more like a drug than anyone had imagined.

In emergency rooms, doctors use a drug called naloxone to stop the effects of heroin and morphine. It blocks opiate receptors in the brain, and a person treated with naloxone gets no high from narcotics.[17] Researchers decided to see what happens when a committed chocolate lover is pre-treated with this same opiate blocker. The results are surprising, and a bit disconcerting. Chocolate becomes not much more exciting than, say, a piece of dry bread. It still fills you up, but it loses its allure. Chocolate's extra pleasure apparently comes from its ability to stimulate opiate receptors in the brain.† The same has been demonstrated for other fat-sugar mixtures, such as ice cream or cookies made with plenty of butter and sugar.[1,18]

---

*Like alcohol, tobacco, and drugs, chocolate's story is quite recent within the broad history of human existence. Cro-Magnon men and women never took a shot of Scotch or a drag off a cigarette, and they never dipped a finger in chocolate syrup, either. Some forms of bitter chocolate were used by Mayans and Aztecs as beverages, but not until the nineteenth century did the smooth, sweetened chocolate-and-cocoa-butter mixtures that are now familiar in candy bars, cookies, and cakes first appear, making it possible to virtually inhale chocolate.

†Some researchers have suggested that the main reason we have natural opiates in our bodies—the endorphins, for example—is to stimulate eating. A bite of food causes the opiate release, which then pushes us to dig in. The pain relief these opiates provide when we are injured may simply be a secondary function.[17]

When researchers offered groups of volunteers trays filled with various snacks, they found that the opiate blocker stopped much of chocolate's appeal. Volunteers ate 90 percent fewer Oreo cookies, 60 percent fewer M&M's, and 46 percent fewer Snickers bars. This was especially true for binge eaters, but also for people without eating disorders. (In case you were wondering, opiate blockers are not generally used in weight-loss programs as, over the long run, they can damage the liver.) In contrast, an opiate blocker leaves the tastes for foods that are not sweet—popcorn, saltines, and corn chips—mostly undiminished.[18]

Don't panic. "Chocolate addiction" does not mean that you will one day rob a gas station to feed your habit. But it does mean that we should look at chocolate cravings as we do any other craving. A food that draws you in like an addiction is likely to be working on specific centers within your brain.

Some food addictions run in families. While most food preferences are not very similar across generations—children learn new tastes from their school chums more often than from their parents—the tastes for hot sauce, black coffee, and beer are remarkably similar between parents and children.[19] Each of these three foods is habituating, to a degree, and a susceptibility to them probably has a genetic basis. We come to like them, apparently, as we associate their taste with the pleasant effects they have on the brain.[1]

What to do? First of all, assess whether a food habit does you any harm. If your chocolate indulgences are rare, there is little reason for concern. It is a mistake to blame a weight problem on sweets if the real problem is in the main dishes that make up your diet. An occasional one-ounce serving of jelly beans, with its one hundred or so calories, is not health food by any means, but a typical four-ounce chicken breast landing on your plate at every lunch and dinner has double the calories and much more fat.

If, however, chocolate or other sweet snacks are adding sizable amounts of sugar and fat to your diet regularly, it may be time for action.

First, take a look at ways to have a chocolate taste that no one will worry about. The recipe section of this book includes desserts made with cocoa powder or carob, which will help you avoid some of the fat and calories of regular chocolate.

Second, see the three-week diet makeover in chapter 7. It is a simple way

THE OPIATE EFFECT OF SNACKING

NO OPIATE EFFECT	MILD OPIATE EFFECT	SIGNIFICANT OPIATE EFFECT
Breadsticks	Chocolate chip cookies	Lorna Doones
Corn chips	Jell-O	M&M's
Jelly beans	Marshmallows	Oreo cookies
Popcorn	Potato chips	Saltines with jelly
Saltines	Pretzels	Snickers bars
Saltines with cheese		

An opiate blocker reduced consumption by 20 to 40 percent for those listed as having a moderate opiate effect and by more than 40 percent for those with a significant effect.[18]

to retrain your tastes without the need to say no to any food forever. This technique is remarkably effective, and you can use it as often as you like.

Third, some people find that magnesium (300 milligrams twice a day) helps. It is available at health food stores and drugstores. Also, the antidepressant bupropion (Wellbutrin) has been shown to knock out chocolate cravings for some people.[20] Bupropion's chemical structure is similar to PEA's.

CARBOHYDRATE CRAVING

Some people crave carbohydrate-rich foods, especially in the winter months. Days are short, and a form of depression called seasonal affective disorder arrives for many people. Cakes, cookies, crackers, bread, and jelly beans seem to act like an antidepressant. The foods can be sweet or starchy; taste is not the issue. The explanation for this phenomenon lies in brain chemistry.

A diet rich in carbohydrates boosts a brain chemical called serotonin, a neurotransmitter that plays a central role in moods, sleep, and other func-

Turn Off the Fat Genes

tions. Serotonin is also the target of common antidepressants, such as Prozac or Zoloft. Carbohydrates work by increasing the passage of the amino acid *tryptophan* into the brain, where it is turned into serotonin. Ordinarily, tryptophan has to compete with all the other amino acids—that is, protein building blocks—for entry into the brain. Carbohydrates spark the release of insulin, which, among its other actions, drives amino acids out of the bloodstream and into the cells of the body. Insulin leaves tryptophan behind, and without all the competition, it slides right into the brain.

In past years, some people believed that the same thing could be accomplished by eating a tryptophan-rich food, such as turkey. However, studies at the Massachusetts Institute of Technology by Richard Wurtman and his colleagues showed that, while turkey does contain tryptophan, it contains even more competing amino acids, with the net result that it actually reduces the amount of tryptophan that gets into the brain. To get a natural antidepressant, the strategy is not to eat turkey but to have a high-carbohydrate diet.

Carbohydrates themselves are not fattening. However, nearly all of them are prepared with loads of fat—shortening, butter, margarine, etc. If a carb-rich diet has contributed to a weight problem for you, the problem is not the carbohydrate itself, which is actually low in calories, but the fatty ingredients or toppings that come along with it: the butter on the bread, the cream sauce on the pasta, etc. The key is to select a diet that gives you the carbs you need, without all the fat. We will address this issue in chapter 8.

NO NEED FOR MORALIZING

Many people feel shame or guilt over anything related to food—as if chocolate lovers should wear a scarlet *C* or eating an extra cookie should lead to the confessional. However, our cravings have more to do with biology than morality. In my work with people hooked not only on foods, but also on alcohol, tobacco, or drugs, I have found it useful to set aside any kind of moralizing or finger-pointing. Instead, we focus on fixing the problem. To paraphrase Jack Nicklaus, you can spend all day trying to figure out how you got your golf ball into the woods, or you can go in there and get it out.

While some psychologists speak of addictive personalities, I do not believe personality has much to do with it. Eating stimulates the release of natural opiates within our bodies. These opiates are relatives of morphine and are powerful motivators. Just as some people are extremely sensitive to the addicting effects of tobacco or alcohol, some people are very susceptible to the natural opiates within their bodies whose release is triggered by foods.[18] If your genes have made you sensitive to the addictive effects of a substance, your vulnerability to it begins with your first taste.

The effect is a bit like smoking. A cigarette may or may not be appealing the first time a person has one. But as its brain effects take hold, a cigarette becomes practically a necessity. Foods can act the same way, as we saw in the previous chapter. Understanding that compulsive eating acts like an addiction, just like smoking, alcoholism, or any other, gives us the power to deal with it.

THE TASTE FOR FATTY AND SALTY FOODS

Some people have a taste for fatty, fried foods. To heck with chocolate. You are looking for fried chicken, a steak, potato chips, or french fries. This taste pattern runs in families. Overweight parents often pass along fatty tastes, along with the all-but-inevitable weight problems, to their children.[1,12,21] The fats in these foods are packed with calories—more than any other food, by far—and easily add to your body fat.[22]

In fact, we like them precisely *because* they are loaded with calories. Our brains are programmed to prefer calorie-dense foods. As with sweets and sugar-fat mixtures, some of the attraction of fatty snacks may come from their ability to stimulate the release of natural opiates inside your body. They are not nearly as powerful in this regard as chocolate, but, as you may have noticed in the table on page 34, potato chips do seem to have a mild opiate effect. In a research study, an opiate blocker reduced their appeal by 35 percent.[17]

Your taste buds are easily fooled. Manufacturers have found that whipping water into margarine or using various proteins to thicken sauces tricks the palate's fat sensors. In general, however, fat substitutes are not a good solution, as they *reinforce* the preference for fatty foods, rather than helping you break it.

Salt is not fattening, although salty foods can facilitate water retention, which can show up on the scale. People would do well to go easy on salt for two other reasons as well.

Salt can raise your blood pressure. Unfortunately, once your blood pressure is up, cutting back on salt is not likely to spare you from the need to take medications. On the other hand, the diet recommendations in chapter 8, along with regular exercise, often have an enormous effect on blood pressure, due both to their nutrient content and their weight-reducing effect.

Less well known, but still important, is that salt contributes to osteoporosis. Any food that is high in sodium (such as the sodium chloride in table salt) encourages calcium from the bones to pass through the kidneys into the urine.[23]

It is surprisingly easy to reduce your taste for salt: for about three weeks simply taper off the amount of salt you use. Canned and snack foods are the worst offenders, so look for low-sodium varieties. To change your taste for salt, also avoid salting the surface of foods, since it is the sodium that touches your tongue that influences your tastes. The goal is not to adjust to the taste of bland foods. Rather, by using less and less salt, your taste for it becomes more acute, and low-salt foods will taste perfectly normal.

TACKLING PREMENSTRUAL CRAVINGS

If you are a premenopausal woman, you may find that cravings are particularly strong during the week before your period.[15] Chocolate sings its siren song especially loudly and may be difficult to ignore.

Think of it as a symptom of "estrogen withdrawal." Throughout the month, your hormones go through dramatic shifts, and during the week before your period, the amount of estrogen in your bloodstream is quickly dropping. Just as a coffee drinker feels out of sorts if he or she has missed a morning dose of caffeine, or a smoker feels jittery when his or her nicotine level drops, a rapid change in estrogen causes symptoms, too: water retention, irritability, mood changes, headaches, and cravings.

Foods can smooth out these hormone changes. In fact, foods actually determine how much estrogen circulates in your blood. Fatty foods increase the amount of estrogen in your blood, and low-fat foods decrease it.

This may sound odd—that somehow the foods you eat affect the amount of hormones in your blood, and that these hormones, in turn, influence cravings. But this is one of the most important health facts to understand.

The hormones coursing through your blood affect a great many things. A persistently high level of estrogen in women is linked to a higher risk of breast cancer. In men, higher levels of testosterone are linked to prostate cancer. Just as fatty foods push your cholesterol level higher, they elevate the amounts of these hormones in your blood. In theory, the more fatty foods you ate during the month, the higher your estrogen level rose as the weeks progressed, and the farther it has to fall at the end of the month, leading to more intense symptoms.

Here is what to do: First, cut the fat. Just as reducing the fat in your diet can make your cholesterol level fall, it will do the same thing for estrogen. If you follow a low-fat diet throughout the month, estrogen does not reach as high a peak as the weeks progress, and *it will not have so far to fall at the end of the month.* Your body will be adapted to that lower level, and withdrawal symptoms—including cravings—are likely to be reduced.

In research studies at the Physicians Committee for Responsible Medicine, we used this kind of diet change to help women reduce menstrual pain and premenstrual symptoms.[24] It does not work for everyone, but many women notice a significant drop in symptoms, including cravings.

This works not simply because the foods are low in fat. When you trade in chicken salad for a pasta-and-bean salad or meat chili for veggie chili, you also get a healthy dose of fiber—plant roughage—from all those whole grains, vegetables, beans, and fruits. The high-fiber diet helps your body eliminate excess estrogens during the month.

Here is how fiber works: The liver filters the bloodstream, removing estrogens and sending them down a small tube, called the bile duct, into your intestinal tract. There, fiber soaks up these excess estrogens and carries them out with the wastes. Grains, beans, vegetables, and fruits have plenty of fiber and keep this removal functioning normally. But if you had tuna, yogurt, an omelette, or any other animal product for lunch, none of these foods have any fiber. Your intestinal tract is missing the fiber it needs to carry estrogens away, and they end up being reabsorbed into the bloodstream.

In addition, high-fiber diets boost a special protein in the bloodstream, called sex-hormone-binding globulin, which binds and temporarily inactivates some of the estrogen circulating in the blood. In general, keeping estrogen levels in the blood fairly modest is good, as estrogen shifts appear to contribute to all manner of problems, including cravings.

As we will see in the next chapter, fiber has other, more powerful actions, as well. For now, the key is that a low-fat, high-fiber combination keeps estrogens from climbing too high during the month, so the change at the

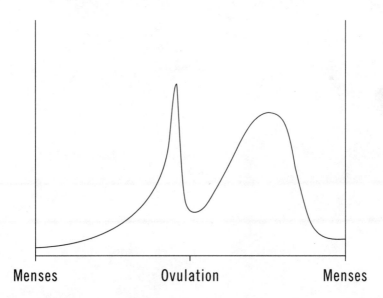

ESTROGEN CHANGES IN THE MONTHLY CYCLE

Menses   Ovulation   Menses

The amount of estrogen in the blood rises and falls dramatically during a woman's monthly cycle. The severe estrogen drop in the week before her period is believed to be the cause of many premenstrual symptoms, including cravings.

end is not so severe. The aim is to smooth out the estrogen roller coaster and avoid the violent hormone shifts that can throw you into cravings.

Even though our tastes are in large part genetically determined, we can choose foods that either work with our tastes or help us change them. In the next chapter, we will look at the genetics of appetite itself—not what we are eating, but how much—and how to tackle an appetite that goes too far.

Chapter Three

Appetite
and the
Leptin Gene

L EPTIN BURST into public awareness on July 27, 1995. Headlines proclaimed that fat mice injected with this appetite-taming hormone slimmed down nearly overnight. It was simple, almost magical. Drug companies salivated over a huge potential market for the wonder drug. They launched armies of researchers into labs to make leptin do for humans what it did for mice.

They failed miserably. In humans, leptin treatments did not do much of anything. In fact, most overweight people already have loads of leptin coursing through their blood. Adding more did not knock out their appetites or make them suddenly svelte.

To return to our earlier analogy of genes sculpting our bodies, appetite certainly has its model in Auguste Rodin, if you will permit me a brief diversion to Paris, circa 1900. Although probably best known for *The Thinker,* the figure that somehow radiates overpowering intensity from the position one might assume while waiting for a bus, Rodin produced hundreds of masterpieces. He was unstoppable, sometimes combining several figures into one statue. His powerful *Victor Hugo* inclines to listen to a winged muse hovering over his shoulder. In his colossal *The Gates of Hell,* Rodin sent countless bodies into a tangle of vices and despair. He had as many as fifty assistants bringing his appetite for the human form to artistic fruition. Meanwhile, his amorous appetites swung from the woman who was to become his wife to a young artist who lived as his mistress and back again, pausing in between to gaze on the nude models strolling around his studio.

Rodin had one advantage you don't have. He could unload whatever he made. Unlike van Gogh, who produced more than 2,000 works in his lifetime and sold exactly one, Rodin shipped monuments to Parisian parks and public buildings and to the museums of Hamburg, Budapest, Prague, Tokyo, Buenos Aires, and Philadelphia, and he gave away busts of his friends.

The human tissue you build is not so easily dispensed with. Chances are, you've been stuck with it for some time. While Rodin could easily chisel away bulges that didn't fit his vision (as he sculpted *Eve,* he compensated for his Italian model's ever-more-apparent pregnancy), you cannot simply carve it away. Even liposuction can't remove more than twenty pounds or so. Many of us have become couch potato versions of *The Thinker,* with our muscularity giving way to softer tissues. It is time to take a closer look at one source of the problem, our appetites.

WHAT LEPTIN WON'T DO, AND WHAT IT WILL

While leptin does not work the same way in humans as it does in obese mice, it is nonetheless important. Here is how it works.

Leptin, made by fat cells, is produced by a gene on chromosome 7. As you gain weight, your expanding fat layer makes more and more leptin, which signals the brain to turn down your appetite. As leptin goes up, appetite goes down. In essence, it provides a feedback system designed to tell the body when it has enough fat. Leptin worked in obese mice because they had a genetic disorder that deprived them of leptin. Overweight humans, on the other hand, have plenty of leptin already—more, in fact, than slim people. Adding modest doses does not help at all. The highest weight loss a team of researchers was able to produce in obese people was thirteen pounds beyond that achieved with a placebo, and that required six months of daily injections pushing the amount of leptin in the bloodstream up to thirty times its normal level, along with a low-calorie diet and exercise. This was better than nothing, but not at all what researchers had hoped for.[1] To the disappointment of pharmaceutical companies, the leptin goose laid no golden eggs.

As this fact emerged, scientists began to speculate that leptin must somehow be malfunctioning in its appetite-taming job. The search for the reasons behind "leptin resistance" continues to this day. In all probability, however, the truth is that high levels of leptin are just not good at stopping weight problems.

Having said that, it is vitally important to keep your own leptin working right. Even though adding extra leptin does not help much, losing the leptin you have can be a serious problem, and unfortunately, that is exactly what happens to many dieters.

Leptin's job is not so much to turn off appetite as to work as part of a switching system that turns it *on* when your food intake drops.[2] The hormone, it seems, is as unceremonious as a ball float in a toilet tank, dropping when the water level falls and opening the pipe to refill the tank. If you lose weight, your shrinking fat stores produce less and less leptin, prompting hunger to return to fill them back up again.[3] Losing just 10 percent of your body fat cuts the amount of leptin in your blood in half.[4]

In fact, you lose leptin even before you lose weight. Just cutting down on food slows down your leptin production. A team of Italian researchers put a group of overweight women on a punishing 300-calorie diet for five days and found the amount of leptin in their blood dropped to half its normal level.[5] Even much less severe diets, say 1,000 calories per day, have a similar effect.[6]

This is the critical take-home message: While an *increase* in leptin is not especially powerful at reducing your appetite or trimming you down, you definitely do not want to court a *decrease* in leptin. It could make your appetite soar. And leptin has other important effects you do not want to lose. It is designed to help you reduce fat storage, improve fat-burning, and maximize your ability to handle sugars.[7]

As we saw in chapter 1, doctors in Cambridge, England, identified two cousins with chromosome 7 abnormalities that blocked the leptin gene. Their appetites were off the scale. Later, the older cousin was treated with leptin-restoring injections, and within a week her meal sizes became smaller and smaller, and she quit sneaking snacks. Over the next year, her weight dropped thirty-six pounds.[8]

In a group of Mexican-Americans, a gene turned up that reduced the amount of appetite-taming leptin they made, resulting in increased body fat.[9] Oddly enough, the misbehaving gene was not anywhere near chromosome 7, where leptin is made. It was on chromosome 2, and it blocked chromosome 7 from making leptin.

If you let your leptin level fall, you are playing with fire, and that is exactly what happens anytime you go on a calorie-restricted diet. You will get a quick drop in leptin and an appetite rebound. Similarly, people with anorexia nervosa, who restrict their food intake to a dangerous degree in pursuit of thinness, have little leptin in their blood. (In contrast, people with bulimia, who eat enormous quantities of food and then purge, have normal amounts of leptin.)[10]

These studies show that your leptin system is disrupted by diets that cut way down on calories. Avoid them. Instead, we will focus more on *what* you eat than on *how much* you eat, as we will see in more detail in chapter 8.

In case you are wondering if you have a leptin deficiency, doctors can easily measure the amount of leptin in your blood. Unfortunately, they may not know what to do with the result. Leptin levels vary widely from one person to the next, and what is "normal" is not entirely clear. In extreme leptin-deficiency cases, other abnormalities are found, including delayed puberty and marked reductions in sex hormones. These are easy for doctors to check.[8]

THE RESTRAINED EATER

In 1975, Peter Herman, a psychologist at Northwestern University in Evanston, Illinois, reported the result of a surprising experiment. He asked volunteers to have a milk shake or two shortly before an ice cream taste test. As your mother might have predicted, the milk shakes ruined most people's appetite, making them eat less ice cream later. However, some volunteers had exactly the opposite response. The milk shake primed their appetite and pushed them into overdrive. They ended up eating more than they otherwise would.

This latter group of people had one thing in common: they had been on diets or were otherwise consciously restricting their food intake.[11] Dr. Herman found that people who were intentionally restraining their eating were set for a binge, and the milk shake was all it took to trigger it.

Diets lead to binges. Let's say you've been dieting all week, living on nothing but low-calorie diet shakes or tiny frozen dinners. As far as your body is concerned, you are starving, trudging through a desert with no food to eat, and if this keeps up, the result could be life-threatening. So your body prepares to take full advantage of any food that comes your way, thinking it might be the last food you'll see for a while. Whether you are aware of it or not, your body gets ready to binge.

At this point, a loved one arrives home with a reward—you can tell what it is just from the frost on the grocery bag: ice cream in your favorite flavor. No, we are not planning to break your diet. This is just a small reward for all your effort. "I'll just have a little bit," you say to yourself. But as soon as the spoon touches your lips, the room goes a bit hazy. You have another spoonful. Then another. You find yourself refilling your bowl. You're eating faster and faster, the room starts spinning around, and before you know it, you're running the spoon around the cracks in the bottom of the carton, tearing it apart to lick the sides—and you've eaten the whole thing.

You then feel overwhelmingly guilty. You scold yourself for your lack of willpower. But the truth is that the binge was not your fault. Your body thought you were starving, and it demanded a binge to save your life. This is called the *restrained-eater phenomenon,* and you cannot do much to stop it. The same thing can happen if you skip meals: you'll almost certainly overeat later in the day.

To prevent your body's natural antistarvation mechanisms from kicking in, follow the Rule of 10: include in your menu each day at least ten calories for every pound of your ideal weight. If you go below this, your appetite will slow down, and a binge can easily follow.

Rather than cut calories, it is much easier over the long run to change the *type* of food you eat. Skipping meals and limiting yourself to minuscule portions is completely unnecessary, not to mention counterproductive.

DISINHIBITION, COMPULSIVE EATING, AND BULIMIA

At times, our best attempts to resist food fail. An obvious time is after having a drink or two. Suddenly, almost *everything* on the menu looks pretty good. More common is the failure of our all-too-weak will when we are bored, lonely, or depressed.

Often it starts with a diet: out of dietary restraint comes a binge. Bingeing may or may not lead to purging, but almost always brings feelings of shame and secrecy. If this has happened to you, remember, it is not a moral failing. The diet did it. And now is a great time to get a handle on it. Here are three steps for tackling the problem:

1. Recognize the problem. Ask yourself these questions:
    *   Is food your usual answer to stress, anger, or sadness?
    *   Do you eat differently when others are around than when you are alone?
    *   Do you hide food?
    *   Do you eat when you are not at all hungry?
    *   Do you snack throughout the day?
    *   Do you order more than one entrée at a restaurant?

    If the answer to any of these is yes, this is not the time to rummage through your childhood memories to understand where the problem came from. For now, we want to get a handle on it.

2. Select the *type* of food you eat from whole grains, legumes, vegetables, and fruits, as described in chapter 8. These foods are very forgiving if you happen to overdo it a bit.

3. Explore other ways of dealing with stress and emotions. Spend time with friends who are not preoccupied with eating, and go places where eating doesn't occur—a play, a class, a museum, etc. I highly recommend Overeaters Anonymous, a support group that meets regularly in nearly every city. OA can help you break out of compulsive eating. It is free and has helped many, many people. Check your local telephone directory or call 505-891-2664 or check the OA Web site, www.overeatersanonymous.org. The combination of a plant-based diet and the support of OA can get you well on your way to coping

with the emotional roller coasters of life without plunging into self-destructive habits, and can even help smooth out the ups and downs a bit.

When you are tackling a weight problem, avoid alcohol and get plenty of rest so your resolve is as strong as it can be. Stock your cupboards and desk drawers with foods that, if you do binge, will not do you any harm. An apple, pear, or rice cake will never have the seductive appeal of a chocolate bar, but if one is there when you need it, it takes the edge off your appetite.

FILLING, NOT FATTENING, FOODS

You can choose foods to tame your appetite. The way not to do this is with fatty foods—salad dressings, cheese, fried foods, etc. They may well kill your appetite, because they slow down the emptying of the stomach. But they also cause the abdominal discomfort you might feel after an especially greasy restaurant meal, and they disrupt the muscular motions that normally guide foods along your digestive tract, like squeezing a tube of toothpaste at both ends at the same time. These fatty foods also pack so much fat, they add the pounds in a hurry.

A different approach uses fiber. You know the old saying about eating at a Chinese restaurant—you are hungry again an hour later? Most Asian restaurants serve white rice, which has been stripped of its natural fiber coat. Without fiber—the "roughage" in vegetables, fruits, whole grains, and beans—the nutrients in white rice are quickly absorbed and hunger soon returns. If you were to make the same meal using brown rice, it naturally makes the nutrient absorption slow and steady, which is not only better for your health but also delays the return of hunger.

Fiber reduces your appetite and helps you lose weight. On October 27, 1999, the *Journal of the American Medical Association* looked at the diets of a large group of people in Alabama, California, Illinois, and Minnesota and found that even modest increases in vegetables, beans, and other fiber-rich foods led to weight loss. People whose diets were richest in fiber weighed eight pounds less, on average, than those with the least.[12]

Many people imagine they will lose weight by eating low-fat yogurt or skinless chicken breast. However, animal products have no fiber at all. A cup of Dannon low-fat yogurt gives you four grams of fat and plenty of sugar, but not a speck of fiber. Chicken breast, fish, egg whites, and all other animal products provide calories, but no fiber at all.

The participants in the *JAMA* study were only scratching the surface of what they could really do. The low-fiber group was getting about ten grams of fiber per day. The high-fiber group doubled this to around twenty. However, you can easily bring this figure much higher. It is not uncommon to see fiber intakes of forty to sixty grams per day in countries where grains, beans, and vegetables are the staples of the diet. Unfortunately, many Asian countries are trading in these traditional foods for hamburgers and fried chicken. Out goes the fiber, and in comes the fat, visible both on their plates and on the scales.

My research team recently ran a study with a group of young women and learned something remarkable about appetite. Their normal calorie intake averaged about 1,900 calories per day. In the study, it plummeted to just over 1,500. Even so, the participants did not become the least bit hungry. How did they do it? We had not prescribed any drugs or even asked them to limit how much they ate. In fact, we encouraged them to eat until they were full and to help themselves to seconds.

The trick was this: we asked them to select their meals from plant sources. Breads, cereals, pasta, vegetables, fruits, beans, peas, lentils, and other foods from plants were unlimited. We asked them to avoid animal products and keep vegetable oils to a bare minimum. They were free to eat as much pasta as they wanted, so long as it was topped with a light tomato sauce, instead of a meat sauce or olive oil. Vegetable soup, lentil soup, black bean soup, gazpacho, or split pea were fine, but creamy soups were out. They could have bean chili, instead of meat chili. A Mexican bean burrito with salsa, a light vegetable stir-fry, and vegetable curries were all fine. It took a bit of getting used to for some participants, since burgers, yogurt, and ice cream were replaced with veggie burgers, soy yogurt, and sorbet. These foods are naturally so low in fat and calories that, even though participants were eating until they were full, their calorie intake dropped, and they lost weight easily.

This kind of change does several important things. These foods are loaded with fiber, which is completely missing from yogurt, chicken breast, and all other animal products. Also, fats were almost completely eliminated. Since fat is by far the most calorie-dense part of foods, when you eliminate it, every bit of food has many fewer calories. So if you happen to take seconds or thirds or fourths, little harm is done. In addition, with no artificial calorie limit, there is no sense of deprivation that might lead to binges later on.

After eight weeks on the diet, we asked the participants to switch back to their normal way of eating for comparison. To our surprise, many refused to do so. They felt good on this new and healthful way of eating. They had lost weight, and many had more energy than they had had in years. Even though they had given up what had once been routine foods, they were glad to have found a healthier way of eating that trimmed their waistlines naturally.

This was not the first study to demonstrate this effect. We had earlier done a study using the same kind of diet for people with diabetes. Their calorie intake dropped, too, by about 300 calories per day, without any increase in hunger. Two-thirds of participants either came off their medicines completely or needed less medicine. Similarly, in Dr. Dean Ornish's well-known program for reversing heart disease, a vegetarian diet caused calorie intake to drop naturally from about 2,000 calories per day to less than 1,800, a drop of 10 percent without even trying to cut calories.

How much fiber should you get in a day? A good rule of thumb is around forty grams. Nutritionists have investigated whether fiber interferes with the absorption of calcium or other nutrients. In fact, even with their generous fiber content, the calcium absorption from most green leafy vegetables is high, over 50 percent, compared to about 32 percent for milk.[13]

Certain kinds of fiber have special effects. *Soluble* fiber, in oats, beans, and fruits, lowers cholesterol levels. *Insoluble* fiber—in wheat, for example—helps keep digestion running smoothly. Many plant foods have a mixture of both types.

There is no need to calculate your daily fiber intake. If you follow the guidelines below, you will get plenty. (One sure sign of inadequate fiber in your diet is the presence of a magazine rack in your bathroom; fiber cures constipation.) Here are three rules of thumb for boosting fiber in your diet:

1. Choose a variety of plant foods (rather than animal products), especially vegetables, beans, whole grains, and fruits. As you'll see in the chart below, topping your oatmeal (old-fashioned oatmeal rather than instant) with soy milk instead of cow's milk brings an extra 3.1 grams of fiber. Animal products present your digestive tract with a large calorie load that has no fiber at all.

2. Keep it simple. The more a manufacturer processes a food, the more it loses fiber. As the chart shows, cereals that are colored and sweetened to appeal to kids have also had most of their fiber removed.

3. Choose foods with natural fiber, not added fiber. For example, oatmeal is naturally fiber rich. On the other hand, cereals that have fiber added at the factory may also have plenty of other additives to make the fiber stick to the product or to provide extra flavoring. Skip them.

WHERE THE FIBER IS—AND ISN'T (IN GRAMS)

FRUITS

Apple (1, medium)	3.7
Banana (1, medium)	2.7
Cantaloupe (1, medium)	1.3
Orange (1, medium)	3.1
Orange juice (1 cup)	0.5
Peach (1, medium)	1.7
Pear (1, medium)	4.0
Prune juice (1 cup)	2.6
Prunes, dried (10)	6.0
Strawberries (1 cup)	3.4
V8 juice (1 cup)	1.0
Watermelon (1 cup)	0.8

BREAKFAST CEREALS

All-Bran (1 cup)	20.0
Cheerios (1 cup)	3.0
Cornflakes (1 cup)	1.1
Froot Loops (1 cup)	0.6
Grape-Nuts (1 cup)	10.0
Oatmeal, instant (½ cup, dry)	2.6
Oatmeal, old-fashioned (½ cup, dry)	3.7
Trix (1 cup)	0.0
Wheaties (1 cup)	3.0

GRAIN AND GRAIN PRODUCTS

Bagel (1)	1.6
Bread, pumpernickel (1 slice)	2.1
Bread, white (1 slice)	0.6
Bread, whole wheat (1 slice)	1.9
Bread, whole wheat pita (1)	4.7
Corn bread (2 ounces)	1.4
Rice, brown (1 cup, cooked)	3.5
Rice, white (1 cup, cooked)	0.6
Spaghetti (1 cup, cooked)	2.4

LEGUMES AND BEAN PRODUCTS

Baked beans (½ cup, boiled)	6.4
Black beans (½ cup, boiled)	7.5
Chick peas (½ cup, boiled)	6.3
Kidney beans (½ cup, boiled)	6.6
Lima beans (12 cup, boiled)	6.6
Peas (½ cup, boiled)	3.5
Pinto beans (½ cup, boiled)	7.4
Refried beans, vegetarian (½ cup)	6.5
Soy milk (1 cup)	3.1
Tofu, firm (1 cup)	5.8

VEGETABLES

Broccoli (1 cup, boiled)	4.6
Carrots (1 cup, boiled)	5.2
Green beans (1 cup, boiled)	4.0
Lettuce, iceberg (4 leaves)	1.2
Mustard greens (1 cup, boiled)	2.8
Potato, baked (without skin)	2.3
Potato, baked (with skin)	4.8
Potato, instant mashed (1 cup)	4.8
Spinach (1 cup, boiled)	4.4

MEATS AND DAIRY PRODUCTS

Beef, ground lean	0
Chicken breast, skinless	0
Halibut	0
Lamb	0
Milk, skim	0
Milk, whole	0
Salmon	0
Yogurt	0

SOURCE: Pennington JAT, *Bowes and Church's Food Values of Portions Commonly Used* (Philadelphia: Lippincott-Raven, 1998).

So what does this mean when you are planning a meal? Let's take an example. If your daily diet consists of roughly 2,000 calories per day, which is typical for an average adult, one meal would supply about 600 calories.

If you were to chose a diner-style breakfast of two eggs, three sausage links, and buttered white toast, you would easily get those 600 calories (643 to be exact), and along with it come 50 grams of fat and 516 milligrams of cholesterol—enough to make a cardiologist blanch. And the meal has all of a half a gram of fiber, from the toast.

If instead you had a half cantaloupe, a large bowl of old-fashioned oatmeal, and whole-grain toast topped with cinnamon, the nutrient balance would look completely different. This breakfast provides a generous 11 grams of fiber, while the calorie content falls to 421, fat content to 8 grams, and cholesterol to zero. The difference in these two breakfasts is that plant products are loaded with fiber and have little fat and no cholesterol, while animal products are just the opposite: plenty of fat and cholesterol, but no fiber.

If, as a snack, you had a cup of yogurt, it would provide no fiber at all, but if you had an apple or pear instead, it would give you a good 4 grams.

Let's have lunch at an Italian restaurant. The pasta-bean soup is a good choice—it's delicious and rich in fiber from the beans and noodles. A mixed green salad adds a bit more fiber (and is low in fat if you keep the dressing on the side) as does a main dish of linguine with a light tomato sauce. If you were instead to eat fish or veal, these foods would fill you up while leaving fiber out.

If you prefer a Latin flavor, a traditional meal of beans and rice with vegetables on the side seems humble, but it is a nutrition powerhouse compared to the hamburgers and french fries served farther north. If you had instead gone to KFC, the fried chicken is loaded with fat and cholesterol, and the only trace of fiber you'll find is in the breaded coating.

For dinner, let's go Chinese: sweet and sour pork has no fiber at all. But all the vegetable main dishes are loaded with it. Or you might like an Indian lentil curry with rice and chapatis. You'll score big, not only for your exotic tastes but also for the natural fiber packed in. At these restaurants, be sure to ask the chefs to minimize the oil used in cooking, or they will likely use the overly large amounts Westerners have come to expect.

These healthful foods satisfy your appetite and are exactly the choices that, in long-term studies, help bring lasting weight loss. They spare you the trouble of low-calorie dieting and the damage it can do to your leptin appetite-control system. Of course, there is much more to the science of weight management, as we will shortly see. But simple changes in the diet set the stage for permanent success.

Whether your leptin system is working well or your chromosome 7 is sound asleep, you can do several things to keep your appetite under control:

1. Low-calorie diets reduce leptin. Avoid them.
2. Having regular meals cuts the tendency to binge.
3. High-fiber meals satisfy hunger with fewer calories.
4. The best high-fiber choices are vegetables, beans, whole grains, and fruits.
5. Regular exercise helps keep the appetite regular, as we will see in chapter 6.

Chapter Four

The Fat-Building Gene

I N   J A P A N , weight problems have always been rare, at least until recently. Where traditional diets, rich in rice and vegetables, remain the order of the day, weight problems are still nearly unheard of. A menu change can put an end to that, however, as evidenced by the emerging epidemics of obesity and heart disease as Western tastes for fast food, meat, and dairy products take hold.

Japanese genes are not changing. But as the diet changes, so do waistlines. Nowhere is this more apparent than in sumo, the centuries-old tradition in which hours of ritual drumming, parading, and salt-throwing are followed by a fifteen-second clash in which one human behemoth throws another out of a ring.

Sumo's most striking competitor in recent years is Akebono, sumo's sixty-fourth *yokozuna,* or grand champion. He has godlike status in Japan, and for good reason. When he enters the ring and slowly rises from his ritual squatting position, he towers over his opponents at 6 feet 8 inches and 500 pounds. Most other wrestlers weigh in at a mere 350 pounds, big enough for the Giants' defensive line, but Akebono tosses them like a salad.

Akebono's popularity comes from more than his tsunami-like strength. As he makes his ritual entry into the ring to open a tournament, raising a massive foot and then stomping it to the floor to drive demons from the ring, then raising his other foot and stomping again to the frenzied shouts of fans in the packed coliseum, he has a tiny glimmer in his eye, betraying the truth about him: Akebono is actually a nice guy. He is as kind and unassuming a mastodon as ever made the earth tremble. When he was promoted to sumo's highest rank, there was no bragging, no bravado. He humbly promised to practice hard and "not to defile the status of grand champion."

Who would have guessed that Akebono, the master of a 1,500-year-old Japanese tradition, is, in fact, an American? Akebono was born Chad Rowan in Hawaii in 1969 of Hawaiian, Irish, Cuban, and Chinese ancestry. Tall and fast on his feet, he went to Hawaii Pacific University on a scholarship to play basketball. But that did not last long. His sports career was quickly shanghaied by Azumazeki Oyakata, a sumo wrestler turned coach, who transplanted him into another world.

How does an eighteen-year-old Yankee hoops player become a 500-pound Mount Fuji? For Chad Rowan, aka Akebono, the answer came in massive servings of chicken, fish, pork, and beef, along with traditional tofu, vegetables, and rice, all washed down with plenty of beer. He gained a massive musculature, a huge fat layer, and a center of gravity akin to that of a Steinway piano. Akebono reached sumo's peak astonishingly fast, rising from novice to champion in just five years. He has also kept his doctors busy worrying about his blood pressure and heart and operating on both his knees, constantly stressed by his oppressive weight.

Chad Rowan's body held exactly the same fat-building genes as Akebono's. A diet change whipped them into action. This is not just a theoretical point. Understanding what activates these genes is the key to blocking body fat. Let us take a look at genetics at work.

CHROMOSOME 8'S KEY GENE

Chromosome 8 holds the gene for a critically important fat-storing enzyme. With the unassuming name of LPL, or *lipoprotein lipase,* it is the switch that either sends fat to be stored on your hips, thighs, and abdomen, or instead burns it for energy. You have probably never heard of LPL, unless you are a metabolism expert. But the most important decision in determining whether you gain or lose weight is not whether to buy this or that piece of exercise equipment, whether to go on a low-calorie diet, or whether liposuction is right for you. The key decision is whether or not to let LPL build fat. If it does, you gain weight. If not, you lose weight.

LPL is all but unknown to dieters and their physicians, but it is the gatekeeper to your fat cells. People struggling to lose weight try all manner of diets, yet few if any realize that this child of chromosome 8 holds the key to weight control.

Until recently, there was little evidence that genes affected how aggressively our bodies store fat. However, we now know that *most* cases of overweight reflect a predisposition to store fat, aided and abetted by eating the wrong foods—or too much food—and getting too little physical activity.[1] German researchers recently found that about one in thirty obese people carries a gene that turns other cells, called fibroblasts, into fat cells, ready to build and store body fat.[2] While this gene does not explain most weight problems, it shows that genes have an effect that is decisive, even if unrecognized.

The gene that makes LPL plays the central role in fat storage for virtually all of us. LPL allows fat to slip from your bloodstream into your fat cells and decides, to a great degree, whether fat will be stored or burned.*

Once your genetic machinery builds LPL, the enzyme takes up its position along the tiny blood capillaries that course through your body fat. It waits on the capillary's inside wall. As particles containing fat flow by in the blood, LPL plucks the fat out. Then, it goes exactly where you don't want it: into your thighs or hips or abdomen for storage as body fat.

The same LPL enzyme is also found in the blood vessels that run through your muscles. There, its effect is very different. Again, it eases fat out of particles in the blood. But as bits of fat pass into muscle cells, they are not stored, at least not for long. Instead, they are used for energy, fueling your muscle movements. In essence, LPL acts like a shovel that moves fat into your cells. In fat cells, you store it. In muscle cells, you burn it.

---

*LPL's name comes from *lipoproteins,* tiny particles carrying fat (lipid) and protein in the bloodstream. The *ase* ending means that lipoprotein lipase is an enzyme—that is, a molecule that speeds up a chemical reaction, in this case, cutting fat molecules loose so they can pass into fat cells.

You can slow down LPL's fat storage. The first step is to cut down its source of raw materials. If you can keep it from getting its hands on fat in the bloodstream, it has less to store.

THE RAW MATERIAL FOR BODY FAT

When a little boy takes a bite of a chicken nugget, his digestive tract quickly separates its fats from its proteins. The fat molecules—of which there is no shortage—pass into his circulation, carried on special particles to the organs and tissues. As they reach the fat tissue, LPL yanks the fat molecules out of these passing particles, so fat can pass into body fat for storage. If he exercises enough to burn fat as quickly as he stores it, he won't become overweight. If not, he'll join the ranks of people struggling with weight problems.

LPL is looking for fat. Fat in foods—not carbohydrate or protein or anything else—provides the principal raw material for building body fat. In 1999, Texas researchers reported on studies of children aged three to seven, measuring their skin folds every year. As is true with every group of kids, some became a bit more plump with each passing year, while others did not. The researchers analyzed their diets and found that the children who put on excess weight were those who had been eating the fattiest foods. It was not necessarily the children who ate the *most* food who gained weight, nor was it the children who ate the most starches. In fact, bread, beans, rice, fresh fruit, and vegetables did not contribute to body fat at all. It was the traces of fat in meat, cheese, fish, salad dressings, french fries, and other fatty foods that found their way into these kids' body fat and stayed there.[3]

If a researcher were to stick a needle into the fatty part of your thigh and pull out a small fat sample, he or she could tell you what kinds of foods you have been eating. The fat you may have on your thighs, abdomen, or anywhere else is made largely of fish fat, chicken fat, beef fat, dairy fat, vegetable oil, or fryer grease—the traces of fat in your diet. It passed from plate to fork, from your digestive tract into your blood, and then—thanks to LPL—into your body fat with very little chemical change. Laboratory tests can examine these fat molecules and trace them back to the fats on your plate eaten months or years earlier.

In one of the first studies to show that body fat comes from food fats,

Turn Off the Fat Genes

rather than other parts of the diet, researchers at a Veterans' Administration home in Los Angeles asked residents to consume either normal diets or diets using vegetable oils instead of animal fats. Then, every four months, each man dutifully had a fat sample removed from his derriere with a tiny needle. Laboratory analysis showed that body fat mirrored the kinds of fats in the foods eaten. Men eating lots of beef and chicken fat ended up with these remnants in their own body fat, while vegetable-fat eaters showed up with vegetable fats stored in their behinds. Some of the men stayed with the experimental diet for as long as five years, and the same pattern held true: the fat they ate is what built fat on their bodies.[4,5]

These studies show that it is not mysterious "excess calories" in candy or cupcakes that build body fat. The issue is much more specific. Carbohydrates did not tend to turn to body fat in these men. There was little trace of any fat made from bagels, spaghetti, or mashed potatoes. Their body fat came from the fats they ate: chicken fat, fish oils, beef fat, and the grease in cheese, butter, shortening, or margarine. These slid into body fat with little change in form. Just as cadmium red from a paint tube becomes rosy cheeks, fat from a chicken thigh or breast adds to your own thighs, hips, and arms. LPL finds it floating by and helps it insinuate itself into your body fat.

It is possible to make fat from the other major parts of foods—carbohydrates and proteins—but your body strongly prefers to build body fat from food fat. And that shows the key to shutting down LPL. (For the role of carbohydrates and proteins in body weight, see chapter 10.)

COUNTERACTING THE FAT-STORAGE GENES

If Akebono's menu were built from rice and vegetables and skipped the fish, poultry, meats, and oils, his LPL would have had little to work with. His knees would not be struggling under such enormous weight; he would be lighter and much quicker on his feet.

A modern analogy for LPL comes from the labor disputes that happen every so often at factories making automobile parts. When a strike occurs, the factory producing a particular part shuts down. And since you can't make a car out of thin air, all the auto factories that rely on those parts have to shut down, too. They stay idle until more parts arrive.

The Fat-Building Gene

The fat factories in your body work the same way. Without raw materials, they cannot build fat. If you avoid fats to the extent you can in the foods you eat, your LPL will be idled, at least to a degree. I say "to a degree" because your body can build fat-protein particles out of other parts of the diet and send them into the bloodstream. However, as we've seen, it is the fat in foods that consistently turns up in body fat.

Recently, a laboratory technician called me to look at a blood sample he had just drawn from one of our research participants. "Look what he had for dinner last night!" the technician said. At the top of the tube was a thick layer of fat floating like a big glob of jelly.

When you eat a Tyson's roasted half chicken breast, your body absorbs every last one of its thirteen grams of fat. Have a serving of french fries, and another ten grams or so follow. The more fat there is in your diet, the more fat passes into your blood. And the more fat in your blood, the more passes into your fat cells.[6]

It is surprising to see what happens when you really get away from fatty foods. I don't mean taking the skin off your chicken and switching to 2 percent milk. I mean really getting away from fats. Having spaghetti with a light tomato sauce, for example. If you've not sprinkled on any cheese, cream sauce, or oil, it has almost no fat at all. Or a bean burrito, made from beans wrapped in a flour tortilla with spicy salsa. Or veggie chili.

Our research studies use these and endless other nearly fat-free choices. And for most research participants, weight loss occurs effortlessly. A typical rate of weight loss is about a pound a week, and as tastes adjust to the lighter foods, this gradual slimming easily continues after the study ends.

Switching from tuna salad to three-bean salad for lunch is painless enough and spares you from the nineteen grams of fat you would otherwise have eaten. Think of it another way: Those nineteen grams of fat pack in 171 calories. It takes a fair amount of exercise to burn them off. The bean salad did it for you without your having to lace up your sneakers.

If you are making spaghetti at home, the choice between a meat sauce and a light tomato sauce seems unimportant. But the tomato sauce cuts out more than five grams of fat per serving, another 45 calories you didn't have to burn off on the treadmill. If you top a baked potato with Dijon

HIGH-FAT VERSUS LOW-FAT CHOICES

(Figures listed are the grams of fat per serving, with the percentage of calories from fat in parentheses.)

WHERE FAT IS		WHERE FAT ISN'T	
Doughnut, glazed	13.7 (51%)	Bagel	1.1 (5%)
Ice cream (1 cup)	14.6 (49%)	Sherbet (1 cup)	3.8 (13%)
Baked potato w/butter	12.4 (34%)	Baked potato, plain	0.2 (1%)
Potato chips (1 oz)	9.8 (58%)	Pretzels (1 oz)	1.0 (8%)
Meat sauce for pasta	6.0 (39%)	Light tomato sauce	0.7 (15%)
Tuna salad (½ cup)	19.0 (45%)	3-bean salad (½ cup)	0.3 (3%)
Turkey frankfurter	8.0 (71%)	Veggie hot dog	0.0 (0%)

SOURCE: Pennington JAT, *Bowes and Church's Food Values of Portions Commonly Used* (Philadelphia: Lippincott-Raven, 1998).

mustard and black pepper (or salsa, ketchup, or any other fat-free topping) instead of butter, you cut out twelve grams of fat, worth 108 calories.

These savings add up to a slimmer waistline. If you are decisive enough about it, the effect is quite rapid.

By the way, you do need some fat in your diet, although nowhere near the amounts most people get. In 1982, researchers reported the tragic case of a six-year-old girl who had been shot accidentally and had lost most of her intestinal tract. She had to be fed intravenously. At first, the supportive treatment worked, but later on she started to develop numbness, blurred vision, and other nerve symptoms and eventually became unable to walk. Her doctors had neglected to provide the traces of essential fats the body needs to function. When they corrected their error, her symptoms soon disappeared.[7] The amount you need is tiny—only about 3 to 4 percent of your calories, which is about one-tenth the amount that most people get. But it's necessary.

If you were to build your diet from vegetables, fruits, beans, and grains,

The Fat-Building Gene

without any added fat at all and without any animal products, your fat intake would be roughly 10 percent of calories, or about twenty to thirty grams per day. Vegetables, fruits, and beans are preferable to grains, in that they are relatively richer in the essential fats that your body needs.

In our research, we ask our participants to weigh and measure the foods they eat. When they focus entirely on the choice of food, and not at all on the amount, their calorie intake drops dramatically. In a group of young active women, their normal calorie intake of nearly 1,900 calories per day dropped to just over 1,500 when we showed them how to cut the fat.[8] We did not ask them to eat less food or to cut calories. But cutting fat cuts calories automatically. Of course, you can increase the volume of food you eat and eventually get close to your previous calorie intake, and in fact, you should do so if you are burning lots of calories through physical activity. But it is hard to avoid a reduction in calories when you rigorously exclude fats. And these simple diet changes keep your LPL from getting its little enzymatic hands on fat it can store on your hips.

In the next chapter, we will go one step further. You can not only make it harder for LPL to find fat to store; you can also shut down LPL itself, to an extent. This is done by controlling the hormone insulin. When you have too much insulin in your blood, LPL activity proliferates in fat cells, busily storing fat, and it diminishes in muscle, where you need it to help burn fat. By controlling insulin, you can, to a degree, control LPL.

FAT CONTENT OF COMMON FOODS

(percentage of calories from fat)

Apple	5%
Asparagus	12
Baked beans (vegetarian)	4
Banana	4
Beef, lean trimmed round	31
Black beans	4
Broccoli	12
Brussels sprouts	12

Carrots	3
Chicken, roasted white skinless	23
Chickpeas	8
Croissant	47
Egg, boiled	61
Green beans	8
Halibut	19
Ice cream	49
Ice milk	27
Kidney beans	4
Lima beans	3
Navy beans	3
Oatmeal	14
Popcorn, air-popped	10
Popcorn, oil-popped	51
Pork, broiled lean sirloin	31
Rice, brown	7
Rice, white	2
Salmon	40
Spaghetti noodles	4
Spaghetti sauce (light tomato)	15
Spaghetti sauce (meat)	39
Spinach	9
Tuna salad	45
Turkey frankfurter	71

SOURCE: Pennington JAT, *Bowes and Church's Food Values of Portions Commonly Used* (Philadelphia: Lippincott-Raven, 1998).

QUIT BLAMING CARBOHYDRATES AND START UNDERSTANDING THEM

Many people mistakenly assume that their body fat comes from carbohydrates—the starches in bread, rice, potatoes, or other foods. Partly, this is due to a premature verdict on fat.

By the 1970s, it was abundantly clear that butter and bacon grease drove cholesterol levels up, contributed to heart attacks, and expanded our waistlines. In response, lower-fat margarines and fat-free snacks appeared on market shelves. Unfortunately, the obesity epidemic did not wane. In fact, it continued to worsen. Some concluded that fat must not have been the problem, and skeptics went so far as to blame their ever-expanding waistlines on fat-free cookies. What they missed, however, was that most people had never stopped consuming 700 to 800 calories in fat from chicken, beef, chocolate, ice cream, and peanut butter every day. These foods contributed as much as ever to weight problems. In fact, people are eating more fat than ever and are simply adding more snacks—fat-free and otherwise—between meals. Fat-free foods only help if they displace fatty foods, not if they are simply added to them.*

U.S. food surveys from 1980 to 1991 showed that the amount of fat Americans ate every day did not drop one iota. It rose. For adults, it averaged eighty-one grams in 1980 and eighty-six grams in 1991. We've added fat-free cookies and sugary sodas to our daily routines, but have never tackled the fundamental problem with the Western diet—its reliance on animal products and other greasy foods that keep LPL busily storing fat.

An average American eats about sixty-four pounds of animal fat and vegetable oil every year, much to the delight of your LPL, which eagerly tucks it under your chin, into your thighs, and all around your middle.[9]

The primary reason that fat is built up in the first place has nothing whatsoever to do with carbs and has everything to do with the high-protein, high-fat foods—meats, dairy products, and eggs—that are traditional in Western countries. There are two important reasons to stop blaming carbs:

First, a gram of carbohydrate has only four calories. A gram of fat has nine.

Second, the thinnest people on the planet live in Asia—in Thailand, rural Japan, China, Vietnam, and elsewhere. They are not reticent about

---

*There is no need to fear fat-free snack foods. In a controlled research study, people given unlimited reduced-fat snacks did not increase their calorie intake at all and the amount of fat in their diets fell. In contrast, those given unlimited access to high-fat snacks increased both their fat and calorie intake.[10]

FAST FOOD, FAT FOODS

	CALORIES	FAT (GRAMS)	% FAT*
BURGER KING			
Cheeseburger	380	19.0	45
Fish sandwich	700	41.0	53
French fries	370	20.0	49
Hamburger	330	15.0	41
Onion rings	310	14.0	41
Pie, apple	300	15.0	45
Shake, chocolate	320	7.0	20
KENTUCKY FRIED CHICKEN			
BBQ baked beans	190	3.0	14
Chicken breast	400	24.0	54
Chicken drumstick	140	9.0	58
Corn bread	228	13.0	51
Garden rice	120	1.5	11
Green beans	45	1.5	30
Mashed potatoes and gravy	120	6.0	45
Red beans and rice	130	3.0	21
MCDONALD'S			
Big Mac	560	32.4	52
Chicken McNuggets	290	16.3	51
Egg McMuffin	290	11.2	35
Filet-O-Fish	440	26.1	53
French fries, 3.4 ounce	320	17.1	48
Hamburger	260	9.5	33
Quarter Pounder	410	20.7	45

*percentage of calories from fat

(continued on next page)

	CALORIES	FAT (GRAMS)	% FAT*
PIZZA HUT			
Meat Lover's pan pizza	340	18.0	48
Pepperoni Lover's pizza	306	14.0	41
Sausage pan pizza	293	15.0	46
Veggie Lover's pan pizza	243	10.0	37
Veggie Lover's pizza	216	6.0	25
TACO BELL			
Bean burrito	420	12.0	29
Bean burrito, minus cheese	398	10.8	24
Burrito supreme	449	18.4	37
Chicken fajita	461	21.2	41
Nachos	322	18.1	51
Soft taco	226	9.7	39
Taco	183	9.7	48

*percentage of calories from fat

SOURCE: Pennington JAT, *Bowes and Church's Food Values of Portions Commonly Used* (Philadelphia: Lippincott-Raven, 1998).

carbohydrates. Traditionally, they have consumed enormous amounts of them in the form of rice, noodles, vegetables, and bean dishes. They ate little meat, virtually no dairy products, and used little in the way of cooking oils. Most people were also physically active. As a result, LPL enzymes throughout Asia had little raw material to work with—there just was not much fat in the diet—and the people of Japan, Thailand, China, and other countries were slim.

The arrival in Asia of fast-food restaurants and Western eating habits has greatly increased the amount of fat consumed in Japan and reduced the use of rice. As the Asian plate looks more and more like its American counterpart, with meaty dishes, dairy products, and so on, Asian waistlines are starting to look more "American," too. Obesity is much more common, as is diabetes, heart disease, and cancer.

The carbohydrates in potatoes, bread, and pasta are simply long chains of natural sugars that provide energy for the body. Some are absorbed quickly, others more slowly. But they all have one nice feature: if you eat more carbs than you need, your body normally does not store them as fat. Instead, your body uses carbohydrates to build special high-energy molecules called glycogen, which are like special batteries packed into your muscles and liver for times when you might need some extra power.

Let's say you were to eat a large amount of bread, jam, and sugared fruit juice—really stuff it in. In a Swiss research study, people did exactly that, eating as much as 480 grams of carbohydrate in a sitting (starchy "complex carbs" in the bread, along with sweet "simple carbs"—that is, sugar—in the jam and fruit juice). This is quite a lot of food, nearly 2,000 calories' worth. But the research team found that virtually *none* of it turned to fat. Where did it go? It recharged their glycogen "batteries" and the rest was burned, producing body heat.[11]

The carbohydrate-glycogen system is like the charger on a shaver, computer, or electric toothbrush. You can run electricity into the charger for hours. After it is fully charged, it simply stops taking on any more electricity and avoids overcharging. Similarly, your body puts the natural sugars from carbohydrates to use building glycogen and can pack a bit less than a pound (roughly 300 grams) of it into your muscles and liver. Once your "batteries" are fully charged, additional calories are released as heat. That's the system. Intentional and repeated overeating of starches could theoretically allow you to put on some weight, but it's not easy.

These same Swiss researchers fed sixteen young men a high-carb diet (80 percent carbohydrate, 9 percent fat, and 11 percent protein) for several days in an attempt to fill up their glycogen stores. They then gave them huge test doses of carbohydrate—2,000 calories' worth—to see if it would turn to fat. But measurements over the next twenty-four hours showed no net fat accumulation at all. Zero. Even people adapted to high-carbohydrate diets generally gain no body fat from a high-carb diet.[12]

Glycogen does not pad your fat stores at all. It adds to your muscles. You will not see much of it on the scale, any more than a couple of batteries in your pockets could weigh you down. Glycogen does hold water—about three times its own weight—but all the glycogen normally in your body, along with the water it holds, weighs only two to four pounds.[11] An athlete could perhaps stuff in enough carbohydrate to double those glycogen stores as an extra energy source for physical exertion, but that is about as much glycogen as the human body can hold. This is obviously not the reason for weight problems.[13] No one ever had a forty-pound glycogen gut spilling out over his belt or glycogen love handles or a glycogen double chin.

PAY ATTENTION TO YOUR SATIETY CUES

Researchers have repeatedly looked for evidence that starches or even simple sugars contribute to fat. They simply can't find it under normal conditions. Researchers conducting these starch-stuffing experiments find that their participants actually have *less* fat (about fifteen grams less) after the meal than before. I am not suggesting this as a way to lose fat, but if your meals don't have any fat in them, your body tends not to build any. Sugars and starches are stored as glycogen for quick energy or burned as heat.

This does not mean we should stuff ourselves with jelly doughnuts. If you push yourself beyond your normal satiety signal—your body's cue telling you you're full—and keep this up day after day, you will gain weight no matter what foods you eat. Researchers at the University of Colorado pushed a group of volunteers to eat 50 percent more than they would normally eat at every meal for two weeks and found that even high-carbohydrate foods can make you gain weight if you ignore your satiety signal and really push it day after day.[14]

If you are going to overeat any kind of food, high-carbohydrate foods, such as rice or pasta, are actually the safest. After recharging your glycogen "batteries" and letting off some body heat, whatever is left over will be stored as fat, but doing so is an inefficient process that burns up nearly a quarter of its calories. On the other hand, fats and oils become body fat easily, packing in almost their full calorie load in the process.

So why are starches—that is, complex carbohydrates, such as potatoes and pasta—so often blamed for weight problems? Two reasons: First, they hang out in dangerous company. A half cup of mashed potatoes has only seventy calories and cannot turn to fat without a difficult biochemical conversion that consumes much of its calorie content. But the tablespoon of butter dripping over the potatoes packs in far more calories, and nearly all of them easily add to body fat. We might blame the potatoes, but it was the butter that caused the blip on the scale.

A plateful of pasta noodles has only about 200 calories and about one gram of fat. If it contributes to overweight at all, it is largely because we use it to hold Alfredo sauce, meat sauce, or olive oil. The pasta is no more the cause of weight problems than the plate underneath it.

The second reason some people blame carbs is that their portion sizes have grown so large that they really are consistently overeating—not just carbohydrate-rich foods, but everything. When I was a child, fast-food restaurants were a novelty, not a part of our daily routine. There were fewer convenience stores than today, and we were not surrounded by candy and potato chips.

Our between-meal beverage was a glass of water. Sodas were a rare indulgence, and they came in small, refillable bottles. Later, they grew to sixteen-ounce bottles, then into twenty-ounce bottles, loaded with sugar. The same portion explosion seems to be happening with everything else. The 7-Eleven convenience-store chain recently eliminated its small coffee cups. The smallest size you can buy now holds twelve ounces of coffee, cream, and sugar, and of course, more caffeine to fuel the agitation of morning commuters. These growing portions do not help. For the most part, however, body fat comes from fat in foods. Take a closer look at the people around you. The folds of fat in their cheeks or arms, their stores of abdominal fat, the fat on their thighs—these came almost directly from fats and oils in the foods they eat. A bit of shrimp fat here, some chicken fat there, some olive oil around the middle.

MAKING FRIENDS WITH YOUR BIOCHEMISTRY

By choosing foods with little or no fat in them, you can starve your fat-storing enzymes—your LPL, that is—of their raw material for building fat. What you are really doing is restoring LPL to its normal function. Your body is not designed to hoard enormous amounts of fat in an ever-expanding waistline. The enzymes designed to liberate the traces of fat from the plants that made up most of the human menu over the millennia have simply been overwhelmed with an embarrassment of fat. Lower-fat foods are much easier for them to deal with.

Let me encourage you not to underestimate the power of simple diet and exercise changes. They can override even serious genetic conditions. Some people, for example, have a rare genetic LPL deficiency in which fat cannot pass into cells and builds up in the blood to such an extent that it actually damages the organs. Using extremely low-fat diets, researchers have found that their clinical problems subside.[15]

We've seen that a gene on chromosome 8 builds the critical enzymes that liberate fat from foods, passing it into body fat. If you deny these enzymes the fatty foods that act as their palette, they cannot produce body fat. In the next chapter, we will see how to go a step further and build or tear down these enzymes themselves by the foods we choose.

Chapter Five

Fat-Burning: Turning the Flame Higher

S O FAR, we have looked at genes that affect our tastes and appetites, and, in the last chapter, the primary gene responsible for packing fat onto our hips and thighs. For most people, attention to these factors, especially keeping LPL less active, solves the problem. They lose weight easily and need go no further.

But if you have a seemingly intractable weight problem, you should know about two more pieces of the weight-loss puzzle. The first piece, genes and your fat-burning speed, we will cover here, and the second, the influence of genes on exercise, is in the next chapter.

Easily the most misunderstood part of our nutrient-managing machinery starts on chromosome 11, where a gene makes insulin, one of the busiest hormones in the body. Many popular diet books have demonized insulin and the carbohydrates that elicit its secretion into the bloodstream. Bread is bad. Rice is fattening. Beware of pasta. The image conjured up is that carbs elicit insulin release, and it, in turn, drives sugars into the cells where they become fat.

The truth is, insulin is your best friend when it is working properly. It does have a devilish side when it gets out of control, but such diet-book descriptions are wildly inaccurate and lead frustrated dieters in completely the wrong direction. When you understand this child of chromosome 11, you'll have an easier time burning calories and getting rid of fat.

Let's start with a look at how your body manages its calorie-burning speed, and then we will see how insulin fits in, not only in keeping your metabolism up, but also for controlling your appetite.

GENES AND YOUR METABOLISM

Each cell in your body is like a tiny foundry, building the girders, supports, and visible features that give your body its form and burning fuels to power all this activity. Your bone cells are constantly tearing down and rebuilding bone. Your skin cells make special proteins and oils to maintain strength and resilience. The cells of your brain and internal organs are busily constructing hormones that control body functions. Before you even open your eyes in the morning, these foundries are hard at work. And they keep working after you hit the snooze button and fall back to sleep.

If you sit perfectly still, perhaps reading or watching television, you might imagine that your body is not burning any calories at all. But feel your pulse. Your heart is pumping. Your liver and kidneys are filtering impurities out of your bloodstream. Your brain is adjusting body functions, conjuring up thoughts and filing away experiences. These organs never rest. Your muscles are ready to move at an instant's notice, so they cannot turn off.

In our research, to measure how quickly an individual burns calories, we use a special apparatus that samples how much oxygen is consumed and how much carbon dioxide the body produces minute by minute. This measurement is called the *resting metabolic rate,* or RMR. It indicates how fast your body at rest burns calories.

RMR is critical to your body weight. It accounts for 60 to 75 percent of all the calories you burn in a day, all without moving a muscle. Another 10 percent comes from the effort required to absorb and digest foods. Physical activity adds the remainder.[1]

If you have a fast metabolism, it means your cells are burning a lot of calories to run these various functions. The more calories they burn, the easier it is to stay slim. If your metabolism is slow, your cells have plenty of leftover calories to store as fat. Turning up your RMR, even slightly, can have a major effect on your weight. If it slows down, weight control is a challenge.

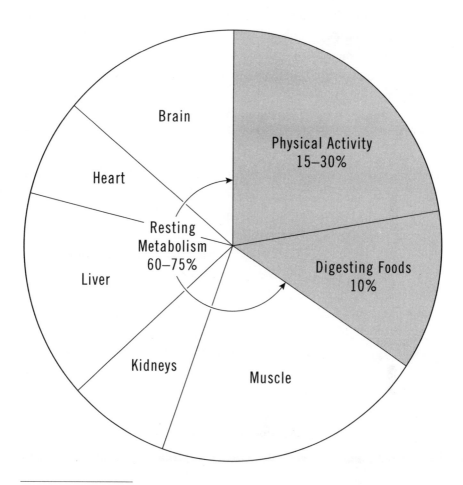

SOURCE: Bouchard C, Dériaz O, Pérusse L, and Tremblay A, "Genetics of energy expenditure in humans," in Bouchard C, ed., *The Genetics of Obesity* (Boca Raton, FL: CRC Press, 1994), 137.

Genes influence your RMR, and they got started before you were born. Studies of identical twins show that one person's metabolism is much more like that of his or her twin (who, of course, shares exactly the same genes) than of another sibling or a parent. And your metabolism is more likely to resemble that of a family member than someone unrelated. If a fast or

Fat-Burning: Turning the Flame Higher   73

sluggish metabolism runs in your family, you may well have inherited that tendency.

Some families have genes for a fast metabolism. The Quebec Family Study found that about one in every fourteen people has a gene for an unusually rapid metabolism, which helps resist weight gain, and this gene, of course, is passed from parent to child, making for a slim family tree.[2] Among Pima Indians, on the other hand, many people carry a gene for a slow metabolism, which contributes to weight gain.[3]

In 1997, a gene with a special effect on RMR was discovered. From its location on chromosome 11 (separate from the insulin gene, which is on the same chromosome), it produces a protein, called *uncoupling protein 2* (UCP2), that disconnects digestion from fat storage so that food you eat is not stored as fat. Instead, energy is dispersed as body heat.[4] People with this gene have higher RMRs and less body fat.[5]

Things slow down as the years go by, whatever your genetic makeup. It's not your imagination. In the Quebec study, mothers' RMRs were 6 percent slower than their daughters', and fathers' RMRs were about 4 percent slower than their sons'.[2] The mothers' and fathers' chromosomes hadn't changed—they still had the same genetic hand they were dealt as life began. But their calorie-burning speed—and yours, too—tends to fall as the years go by. Mostly, this is because inactivity has led to a loss of muscle tissue. If you keep your youthful musculature, your metabolism will tend to stay young as well.

As powerful as genes are, they explain only about 30 to 40 percent of the difference in metabolism from one person to the next. In other words, your metabolism depends more on diet, exercise, and other factors that you can potentially control, regardless of your genetic inheritance.

If you are not one of the lucky ones with a turbocharged chromosome 11, you still have a way to boost your metabolism. In fact, you *are* affecting it several times a day without realizing it, simply by the foods you select.

THE DAMAGE OF DIETING

First, let me show you how *not* to do it. Some ways of changing your metabolism you will want no part of. Surprisingly enough, gaining weight increases your RMR. Body fat and the extra muscles that support it need calories to sustain themselves. The more weight you gain, the more calories your body actually burns. Unfortunately, the increase in your RMR that comes from adding extra weight is not enough to melt away the extra fat, but it does tend to keep your weight from climbing endlessly. Sooner or later, you'll reach a plateau.

When you lose weight, your calorie-burning speed goes in the opposite direction. Just as a Ferrari takes less gas than a Mack truck, your newly slimmed body needs fewer calories to keep its parts running.

Here are the numbers: for every three pounds of fat you lose, you'll also lose another pound of lean body tissue, mainly from muscles that were used to hold up that extra fat. So if you lost twenty pounds, fifteen pounds of it were fat and five pounds were muscle or other supporting tissues. Without those calorie-burning muscles, your body burns fewer calories each day.

Your body burns about nine calories less each day for every pound you lose (or twenty calories less per kilogram lost). So if you lose twenty pounds, your body will burn 180 calories less each day. Hopefully, your hunger will drop similarly. If it doesn't and you eat as much as before, your weight will likely climb back up.[6]

Exercise can preserve your muscle mass and prevent part of this drop, although some decrease in calorie-burning is inevitable if you trim down significantly. More on exercise in chapter 6.

Very low calorie diets cause a severe metabolism drop—beyond that which you would expect simply from being in a smaller body. It is as if your body senses that you're scrimping on food and automatically burns calories more slowly. Within just a day or two of starting a diet, your built-in antistarvation defenses go to work, and it becomes more and more difficult for you to burn calories. If you have any genetic tendency toward a slow metabolism, a low-calorie diet makes things worse.

A blood sample would show what your body is doing. It has cut way back on its production of two compounds—thyroid hormone and norepinephrine—that your body uses to maintain your metabolism. With less of these in your blood, your RMR drops.[7]

Researchers at the University of Pennsylvania put a group of patients

HOW A DIET DAMAGES YOUR METABOLISM

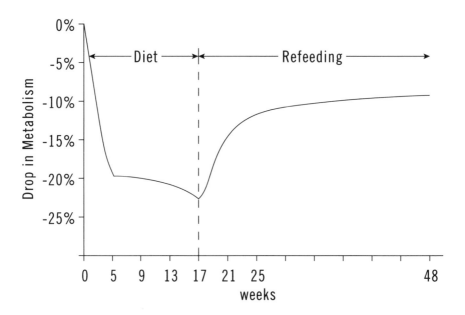

A low-calorie diet causes a marked lowering of the body's calorie-burning speed, which takes several weeks to return to normal after the diet ends.

SOURCE: Wadden TA, Foster GD, Letizia KA, and Mullen JL, "Long-term effects of dieting on resting metabolic rate in obese outpatients," *JAMA* 264 (1990): 707–11. Copyrighted 1990, American Medical Association.

on a 420-calorie liquid diet. Their RMR plummeted an average of 20 percent in the first month.[8] Researchers at Rockefeller University found a similar effect in a group of seriously overweight people. They fed each of them an 800-calorie formula diet until they lost 10 percent of their body weight, which took from six to fourteen weeks. In the process, their average daily calorie-burning had slowed down by 18 percent, or 550 calories less each day.[9] Researchers at the University of California at Los Angeles found the same thing. Even vigorous exercise could not counteract the diet's tendency to slow down their metabolic rates.[10]

Here is the worst part: when you come off a diet, *your metabolism remains slow for several more weeks.*[8] It is as if your body is worried that starvation could recur, so it keeps a slow metabolism to pack a little extra fat on your thighs, just in case you need it later. This metabolism-slowing causes the infamous "yo-yo" phenomenon, in which a diet helps you lose a bit of weight, but it also slows your RMR, so you rebound to a weight higher than you started with. Another diet follows, leading to some weight loss but also more slowing of your metabolism, causing weight to climb again.

INSULIN AND THE AFTER-MEAL BURN

You can rev up a slow metabolism, and this is where insulin comes in. Think of insulin as a worker in a bronze foundry. Like a sweat-covered laborer heating bronze to 2,000 degrees, and then carefully pouring the white-hot metal into a ceramic mold where it will cool and become a statue, insulin's job is to push proteins and sugars into the cells of your body to build body parts and energize your movements. These nutrients come from the foods you eat. They trigger your pancreas to release insulin into your bloodstream, made according to chromosome 11's specifications. Insulin travels to your muscles, liver, and fat tissues, where it pushes proteins and sugars into your cells.

In the process, your metabolism rises dramatically. The reason is that, inside the cell, these natural sugars are building glycogen, the quick-energy "batteries" we learned about in chapter 4. Building glycogen is a big job, causing your cells to actually release calories in the form of heat, just as factory workers making bronze sweat off plenty of calories. This is called the *thermic effect of food,* or TEF. It's a nice way to burn calories. All you do is eat, and your body does the rest. These calories are gone forever—they never even get a chance to turn into fat.

This after-meal metabolism boost depends on two things. First, some foods spark a terrific burn, while others provide none at all. Glycogen is made from natural sugars that come from carbohydrates. So spaghetti delivers a nice after-meal burn. So do most vegetables and beans. They are glycogen builders. The foods that cause the best burn are those that contain plenty of complex carbs (e.g., pasta, rice), or foods containing both carbs and protein. For example, broccoli and other vegetables are about 50 percent complex carbs and 40 percent protein, a mix for a good burn.

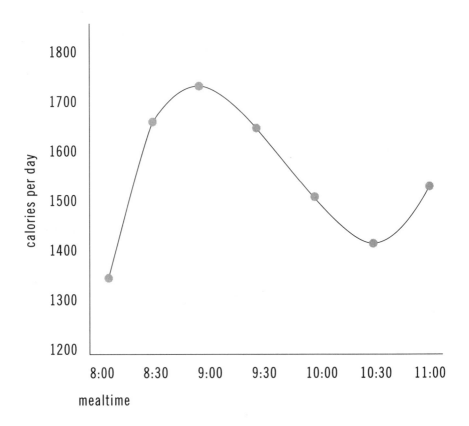

THE AFTER-MEAL BURN

This graph, from one of our research volunteers, shows the typical rise in metabolism after a carbohydrate-rich meal, called the thermic effect of food. The metabolism stays high as foods recharge your glycogen stores.

On the other hand, butter, chicken grease, and olive oil are just fat, and they deliver a much poorer burn. In an experiment, Swiss researchers fed seven young men high-fat meals, with fat totaling 52 percent of their calories, roughly the amount found in fried chicken or french fries. They looked for any evidence that the men's bodies would respond by burning off some of this excess fat. No dice. Excess fat is mainly stored. It goes from

Turn Off the Fat Genes

fork to lips to your body fat. Replacing healthy complex carbs with a load of fat, as in this study, robs you of much of your after-meal burn.[11] So the first key is to choose foods that supply the right raw materials.

Second, the after-meal burn depends on insulin getting nutrients into your cells.[12] This overworked foundry laborer has an important job to do, and unfortunately some people poison their insulin. By eating fatty foods, or by accumulating a fair amount of body fat, insulin is awash in grease. The cell's tiny machinery that is supposed to convey nutrients inside becomes resistant to insulin, as if insulin's hand is slipping on a greasy doorknob. The pancreas then responds by sending more and more insulin into the blood to try to get proteins and sugars into the cells. The ever-increasing amounts of insulin can be seen on blood tests, and this is when insulin becomes a problem.

GLYCOGEN

WHEN INSULIN MISBEHAVES

You do not want to let insulin work overtime. When it does, it has a couple of features that do not help. Its job, as you recall, is to push nutrients into your cells. As it does so, *it temporarily shuts down your fat-burning machinery.* Insulin operates on the theory that it has been sent into the blood because you've got plenty of nutrients from the meal you have just eaten. That being the case, it stops the release of fat from your body fat stores to keep it in reserve for some future time when food is in short supply.

In the short run, this is no problem. Insulin pushes sugar and protein into the cells and then goes away. The interruption in fat-burning is brief. But when insulin resistance develops, the body produces more and more insulin, and it shuts off fat-burning more effectively than it should.

Imagine if your foundry workers were loaded up with bacon, doughnuts, and other fatty foods. Flabby and out of shape, with their greasy fingers slipping all over the machinery, they struggle to do their jobs. The manager sends more workers over to push nutrients into the cells, and they neglect the jobs they are supposed to be doing.

A persistent insulin rise actually increases LPL in your fat tissues, with the result that you can store fat more and more easily, and decreases LPL in your muscles, impairing your ability to burn fat.[13,14] This means that every speck of fat in your blood will find more LPL eager to stuff it into your fat tissues, and less LPL able to put it into your muscles to burn.

Your insulin might be overworking for another reason. In addition to insulin resistance, in which extra insulin pours into the blood to make up for the cells' intransigence, insulin will also overwork if you never stop eating. An endless stream of cookies, sodas, and other foods never gives insulin a chance to rest. If you have a constant supply of snacks, your body has no need to use its fat, and insulin keeps your fat-burning processes slower than they would normally be.

A lack of fiber adds to the problem. Normally, fiber—plant roughage—helps keep insulin levels in check by slowing the release of sugars from the foods you eat, among other functions. When researchers track people's diets and measure insulin in their blood, those with the most fiber have the lowest insulin levels.[15] But many people are missing out on fiber because, as we saw in chapter 3, meats, eggs, and dairy products have no fiber at all, and refined foods, such as white bread, have had most of their natural fiber removed. Candy bars, sodas, milk, and ice cream fill us up, but have no fiber at all.

So how do we get this metabolic boost? The best foods for an after-meal burn are those that are

- high in healthy carbohydrates to build glycogen
- high in fiber to keep insulin on track
- low in fat to make insulin efficient

This means that the best choices for building glycogen, while keeping insulin efficient, are

Beans, peas, and lentils of any variety
Fruit—e.g., apples, bananas, oranges, pears
Green vegetables—e.g., asparagus, broccoli, spinach
Pasta (despite having lost fiber, it releases its sugars slowly, as noted below)
Sweet potatoes or yams
Whole grains: brown rice, barley, bulgur, pumpernickel bread, or oatmeal

Poor choices include

Dairy products and eggs—they have no fiber and most are high in fat
Foods with added fat, such as french fries and salad oils
Olives, nuts, seeds, and avocados, as they are among the few plant foods that are naturally high in fat
Poultry, beef, or other meats—they have no fiber or complex carbohydrates and have more fat than you need
Sugar candies and sodas—lots of sugar, but no fiber
White bread, and most other refined grain products, as they have lost most of their fiber

THE SPECIAL EFFECT OF FRUIT

Fruit gives you a special benefit. Its natural sugar, fructose, delivers an especially good burn. It works even if your insulin is not working well because fructose is rapidly metabolized in the liver, and it enters these cells without any help from insulin at all. So while you are waiting for your insulin function to improve, you can get a great burn right now from bringing plenty of fresh fruit into your diet.[16]

CARBOHYDRATES ARE YOUR FRIENDS

Let's avoid a couple of common misperceptions. In and of itself, insulin is not a bad guy at all. It only becomes a liability when it is thrown into overdrive. This is usually caused by too many greasy foods impairing insulin's ability to work and forcing your body to make more and more of it to compensate, or by a never-ending stream of snacks forcing insulin into constant action.

Some popular diet books have naively suggested that the way to deal with malfunctioning insulin is to avoid the carbohydrates that elicit its release. Cookies, bread, pasta, and rice are verboten on these diets. While there is some value in being choosy about the *type* of carbohydrates you favor, as we will see shortly, the real answer is to replace fatty foods with healthier foods, rich in complex carbohydrates and fiber, so your insulin can start working properly again. Rather than avoid carbs, it is better to fix your body's natural ability to process them.

It is worth remembering that carbohydrates have only four calories per gram, while fats have nine. Carbohydrates are not the enemy. They are, in fact, our natural energy source. In our studies, our research participants boost the fiber and cut the fat in their diets, which causes powerful weight loss in and of itself. Rather than avoid carbohydrate-rich rice, pasta, and vegetables, they are free to eat as much of these foods as they like. In the process, they quickly get their insulin working better, which supports an easy and often dramatic weight loss.

Turn Off the Fat Genes

Participants in one of our recent studies had adult-onset diabetes, a condition in which insulin is failing to get sugar into the cells. As sugar builds up in the blood, some of it passes through the kidneys and into the urine, where we detect it on laboratory tests. To get their insulin working better, we asked our participants to eliminate animal products, which, of course, eliminates all animal fats. We also asked them to keep vegetable oils to a bare minimum. As the weeks went by, most participants' insulin began to work better and better, so they were able to cut back on their medicines, and the amount of sugar in their blood dropped by about 30 percent. Although we used no calorie limit at all, the average participant lost sixteen pounds in just twelve weeks.

TAKE YOUR INSULIN OUT FOR A WALK

In the study mentioned above, we did not use an exercise regimen because we were investigating the effects of diet alone. However, exercise gives insulin an extra boost. For reasons that have never been clear, exercising muscles love to take sugar out of the bloodstream. It is as if your muscles know their glycogen "batteries" are being drained by the work they are doing, so they draw in sugar to recharge them.

Using a twenty-two-day program, researchers at Laval University in Quebec found that exercise improved insulin sensitivity dramatically. Peak insulin levels after meals dropped by more than 20 percent, meaning that the body needed less insulin to do the job.[17] The researchers used an exercise bike, but any kind of aerobic exercise will do: walking, step aerobics, dancing, running, or any other.

YES, IT REALLY WORKS

You can, in essence, rehabilitate a sluggish calorie burn. Focus on two things: Have plenty of glycogen-building foods, which means those high in complex carbs and fiber. These give you a good after-meal burn. And cut the fat to keep your insulin efficient at pushing nutrients into cells without slowing down your fat-burning unduly.

It works. Researchers fed a high-carb, very low fat diet (80 percent carbohydrate and 9 percent fat) to a group of young men. They found that their after-meal burn in response to a test meal improved substantially in a matter of days—rising about 60 percent higher than for those who had been on a high-fat diet.[18]

Your success will build on itself. As you lose body fat, insulin gets more and more efficient, and you get a better and better after-meal burn. Researchers from Columbia University found that a group of overweight men, weighing about 220 pounds (101 kilograms) on average, had an after-meal burn of about 4 percent of the calories in a test meal, while men at their ideal weight got nearly a 10 percent burn.[19] This gradually improving TEF burn that comes as you slim down helps prevent your weight from bouncing back up after you lose it. But don't wait for it. Choose the foods that push your burn now.

GENES AND TEF

It will not surprise you to learn that your after-meal burn, or TEF, is influenced by genetics. Identical twins, who, of course, have the same genes, get almost the same response from a test meal. Fraternal twins, whose genes are no more alike than those of any other siblings, are a bit different in their after-meal burn.[20] And when a pair of twins begins an exercise or diet program, the changes in their measured TEF are similar to each other, while another twin pair might get a greater or lesser effect.[17]

Genes also influence how quickly and easily you can turn up your insulin. In a research study at Louisiana State University, a group of women were asked to cut down on the fat they ate, limiting it to about 20 percent of calories, which is about one-third to one-half less than most Americans get. The change promptly boosted their insulin sensitivity. Nonetheless, the change provided a bigger benefit for some women than others. The improvement in insulin sensitivity was 6 percent for African-American women, but 20 percent for Caucasian women, for reasons that are not clear.[21]

This does not mean that any given individual will get a larger or smaller benefit, because we all vary enormously and these numbers are simply averages. The food guidelines in chapter 8 will bring you well beyond the diet the Louisiana researchers used. Virtually *everyone* can pump up insulin sensitivity by getting away from animal products and their load of fat, keeping vegetable oils to a bare minimum, and exercising regularly.

POWER PLANTS

The more your diet relies on plant foods, the better, as researchers at the University of Vermont found in the late 1980s. They checked the resting metabolic rates of twenty-three healthy young men, twelve of whom were vegetarians while the remainder were nonvegetarians. The vegetarians were slimmer, weighing in at an average of 157 pounds, compared to 173 pounds for the nonvegetarians. This was no big surprise. Vegetarians are usually slimmer than meat eaters. However, because the nonvegetarian men had more body fat and also greater fat-free weight, they ought to have had a higher metabolism. But the surprise was that the vegetarians' metabolic rates were actually a touch higher (regardless of whether it was measured per unit of body weight or per unit of fat-free weight).[22] Something about their adopted diet kept their flame turned up a notch.

WHEN FOODS ARE REALLY APPETIZERS

Overworking your insulin presents one additional danger. It can affect your appetite. Perhaps you have noticed that your refrigerator becomes magnetic in the evening, inexplicably attracting you even though you had finished dinner not long before. The reason may lie in the dinner itself, or even in a meal earlier in the day.

Certain foods stimulate the appetite. Specifically, foods that make your blood sugar rise and fall abruptly draw you back for more. In contrast, foods that release their natural sugars gradually keep your appetite at bay. Here is what is happening:

Your body's basic energy source is found in long molecular chains of complex carbohydrates found in the starchy parts of vegetables, beans, grains, and fruits. When you digest these foods, the carbohydrate chains release their sugars one by one to keep your brain and other organs working.

Some foods release their sugars slowly. Beans or lentils, for example, do so over a matter of hours.[23] This is good. Your body responds by releasing small amounts of insulin to escort these sugars from the bloodstream into the cells of the body where they can be put to use. This slow, steady fuel supply gives you the energy you need, and as a result, you feel satisfied for several hours.

CHOOSING FOODS FOR AN AFTER-MEAL BURN

Foods that promote an after-meal burn are those rich in healthy carbohydrates and fiber, shown in the first two columns. Also, they should be low in fat, preferably with a low glycemic index (GI). There is no established GI cutoff, but a value over 90 would generally be considered high.

How do you quickly size up a food? Check the fiber:fat ratio, a rough but handy guide. Generally, you are looking for numbers above 2. A dash indicates figures that do not apply or have not yet been ascertained.

	CARB (g)	FIBER (g)	FAT (g)	FIBER:FAT	GI*
FRUITS					
Apple (1, medium)	19.0	2.4	0.4	6.0	52
Apple juice (1 cup)	29.0	0.2	0.3	0.7	58
Banana (1, medium)	26.7	2.7	0.5	5.4	76
Grapefruit (½, medium)	9.9	1.4	0.1	14.0	36
Grapes (1 cup)	15.8	0.9	0.3	3.0	62
Mango (1, medium)	35.2	3.7	0.6	6.2	80
Olive (1, medium)	0.3	0.1	0.4	0.3	—
Orange (1, medium)	15.2	3.1	0.1	31.0	62
Orange juice (1 cup)	25.8	0.5	0.5	1.0	74
Peach (1, medium)	9.7	1.7	0.1	17.0	40
Pear (1, medium)	25.1	4.0	0.7	5.7	51
Pineapple (1 cup)	19.2	1.9	0.7	2.7	94
Watermelon (1 cup)	11.5	0.8	0.7	1.1	103
GRAIN PRODUCTS					
Angel food cake (1 oz)	16.4	0.4	0.2	2.0	95
Bagel (1)	37.8	1.6	1.1	1.5	103
Barley, pearled (1 cup)	44.3	6.0	0.7	8.6	25
Bread, pumpernickel (1 slice)	15.2	2.1	1.0	2.1	58

*GI uses white bread as reference value of 100.

	CARB (g)	FIBER (g)	FAT (g)	FIBER:FAT	GI
Bread, rye (1 slice)	15.5	1.9	1.1	1.7	92
Bread, white (1 slice)	12.4	0.6	0.9	0.7	100
Bread, whole-meal (1 slice)	12.9	1.9	1.2	1.6	99
Bulgur (1 cup)	33.8	8.2	0.4	20.5	68
Cereal, All-Bran (1 cup)	46.0	20.0	2.2	9.1	60
Cereal, Cheerios (1 cup)	23.0	3.0	2.0	1.5	106
Cereal, cornflakes (1 cup)	26.0	0.0	0.5	0.0	119
Cereal, oatmeal (1 cup, cooked)	25.3	4.0	2.3	1.7	87
Corn chips (1 oz)	16.1	1.4	9.5	0.1	105
Popcorn, air-popped (1 oz)	21.8	4.2	1.2	3.5	79
Spaghetti (1 cup)	39.7	2.4	0.9	2.7	59
Spaghetti, al dente (1 cup)	39.7	2.4	0.9	2.7	52
Rice, brown (1 cup, cooked)	45.8	3.5	1.6	2.2	79
Rice, parboiled (1 cup, cooked)	44.5	0.6	0.4	1.5	68
Rice, white (1 cup, cooked)	44.5	0.6	0.4	1.5	81

LEGUMES

	CARB (g)	FIBER (g)	FAT (g)	FIBER:FAT	GI
Baked beans (vegetarian, ½ cup)	26.1	6.4	0.6	10.7	69
Black beans (½ cup)	20.4	7.5	0.5	15.0	43
Black-eyed peas (½ cup)	16.8	4.2	0.3	14.0	59
Chickpeas (½ cup)	27.2	5.3	1.4	3.8	47
Kidney beans (½ cup)	20.2	6.6	0.5	13.2	42
Lentils (½ cup)	20.2	7.8	0.4	19.5	41
Lima beans (½ cup)	19.7	6.6	0.4	16.5	46
Navy beans (½ cup)	24.0	5.8	0.5	11.6	54
Peanuts (1 oz, dry-roasted)	6.1	2.3	14.0	0.2	21
Peas (½ cup)	10.7	3.5	0.3	11.7	56
Pinto beans (½ cup)	22.0	7.4	0.5	14.8	64
Soybeans (½ cup)	8.6	5.2	7.7	0.7	25

(continued on next page)

	CARB (g)	FIBER (g)	FAT (g)	FIBER:FAT	GI
VEGETABLES					
Asparagus (1 cup, boiled)	7.6	2.8	0.6	4.7	—
Broccoli (1 cup, boiled)	7.8	4.6	0.6	7.7	—
Carrots (1 cup, boiled)	8.2	2.6	0.1	26.0	101
Potato, baked	51.0	4.8	0.2	24.0	121
Potato, new	—	—	—	—	81
Potato chips (1 oz)	15.0	1.3	9.8	0.1	77
Spinach (1 cup, boiled)	7.8	4.4	0.4	11.0	—
Sweet potato (1, baked)	27.7	3.4	0.1	34.0	77
Yam (½ cup, baked)	18.8	2.7	0.1	27.0	73
SWEETS					
Chocolate (0.5 oz)	8.2	1.0	4.5	0.2	70
Honey (1 tablespoon)	17.3	0.0	0.0	0.0	104
Jelly beans (1 oz)	26.1	0.0	0.1	0.0	114
Life Savers (2 pieces)	5.0	0.0	0.0	0.0	100
Sugar (sucrose, 1 teaspoon)	4.0	0.0	0.0	0.0	92
MEATS, DAIRY PRODUCTS, AND EGGS					
Beef, trimmed round (3.5 oz)	0.0	0.0	6.0	0.0	—
Chicken breast, half, skinless	0.0	0.0	3.1	0.0	—
Egg, boiled	0.6	0.0	5.3	0.0	—
Halibut (3 oz)	0.0	0.0	2.5	0.0	—
Ice cream (½ cup)	15.6	0.0	7.3	0.0	87
Ice milk (½ cup)	15.0	0.0	2.8	0.0	—
Pork, lean sirloin (3.5 oz)	0.0	0.0	6.7	0.0	—
Salmon (3 oz)	0.0	0.0	6.9	0.0	—
Tuna salad (½ cup)	19.3	0.0	19.0	0.0	—
Turkey frankfurter	0.0	0.0	8.0	0.0	—

SOURCES: Pennington JAT, *Bowes and Church's Food Values of Portions Commonly Used* (Philadelphia: Lippincott-Raven, 1998); and Foster-Powell K and Brand Miller J, "International tables of glycemic index," *Am J Clin Nutr* 62 (1995): 871S–93S.

A slice of white bread behaves differently. Its starches break apart quickly, and sugar rushes into the bloodstream. Insulin tries desperately to shove this influx of sugar into the cells of the body and will likely overreact, pushing too much sugar out of the bloodstream. The result is a rapid rise and fall of blood sugar. And that is where trouble starts. This rapid rise and fall signals you to eat again, and your refrigerator suddenly becomes irresistible.

It is reasonably easy to divide foods into those that release sugars slowly and those that release them quickly. The best choices for a slow, steady energy release are beans, vegetables, fruits, and certain grains and grain products: pasta, barley, bulgur (cracked wheat), and whole-grain bread, such as pumpernickel. In contrast, refined grains (e.g., white bread), potatoes, and some sweeteners, such as honey, release their natural sugars more rapidly.

Some foods are surprises. Fruit sugar (fructose) has much less effect on your blood sugar than white bread.[24] And even though white bread and spaghetti are both made from white flour, spaghetti releases its sugars more gradually, with a blood sugar peak only about half as high as that caused by white bread.[25] Parboiling rice or other grains has the same beneficial effect.[24] Long-grain rice seems to release sugars more slowly than short-grain rice.

Baked goods can release sugars quickly or slowly depending on how they were made. Old-fashioned breads made from coarsely ground flour release their nutrients gradually, while most modern bakeries pulverize wheat into a fine powder that is rapidly digested.[24]

A new potato releases its sugars fairly slowly, but a russet potato does so quickly, faster even than white bread. The difference is probably in the more glutinous texture of the new potato, in contrast to the more powdery russet.[26]

As you might guess, a banana yields its sugars faster as it ripens, and in general, the more you cook a food, such as pasta or rice, the more quickly it releases its sugars. This may explain the remarkable weight loss often seen in people who consume large amounts of raw foods: melons, cantaloupes, cucumbers, carrots, apples, oranges, etc. These foods may contain substantial amounts of natural sugars, but they release them only gradually.

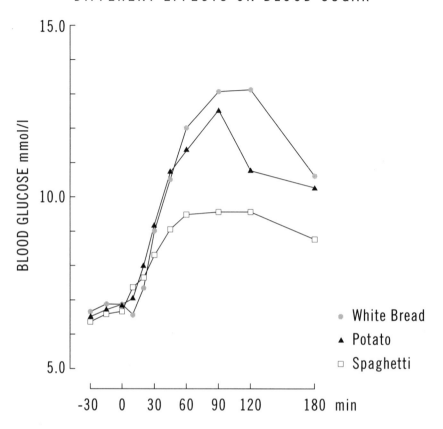

DIFFERENT EFFECTS ON BLOOD SUGAR

- White Bread
- Potato
- Spaghetti

Even though both are made from wheat flour, spaghetti causes much less change in blood sugar than white bread, due to the effects of processing. A potato is similar to white bread in its effect on blood sugar.

SOURCE: Rasmussen O, Winther E, Arnfred J, and Hermansen K, "Comparison of blood glucose and insulin responses in non-insulin-dependent diabetic patients: Studies with spaghetti and potato taken alone and as part of a mixed meal," *Eur J Clin Nutr* 42 (1988): 953–61.

Turn Off the Fat Genes

Nutrition scientists measure the effect various foods have on blood sugar and record the result as the *glycemic index*. Unfortunately, differences in how food is selected and prepared have kept scientists arguing over the importance of the glycemic index. The best evidence shows, however, that the effect foods have on blood sugar really can make a difference on appetite, and also in the regulation of cholesterol and blood fats.

Researchers at Boston's Children's Hospital gave teenage boys various breakfasts, then watched their snacking habits later in the day. Children served regular oatmeal had a slow rise and fall in blood sugar and did not snack much later in the day. But those who ate instant oatmeal had a rapid rise and fall of their blood sugar and became hungry soon after the meal. In fact, they took in 53 percent more snack calories than those who had regular oatmeal for breakfast.[27]

Most brands of instant oatmeal are heavily processed and cause a rapid release of sugars. On the other hand, if your breakfast consisted of a big bowl of "old-fashioned" oatmeal, made from fiber-rich rolled oats, its natural sugars are released gradually. There is no dramatic rise and fall of sugar in your blood, no big insulin response, and no tendency for hunger to return right after breakfast. Other high-fiber cereals work the same way. Bran cereal, for example, releases its sugars gradually, keeping hunger at bay more effectively than cereals from which fiber has been removed.[28]

The same holds true at dinner. Swiss researchers found that shepherd's pie made with beans kept the appetite at bay for hours. When the recipe was changed to replace beans with potatoes, hunger returned sooner, within about an hour for some people.[29] Beans release their natural sugars slowly, while potatoes release theirs quickly. The result is a very different effect on how soon you're ready to eat again.*

The influence of a meal extends beyond how soon you'll want the next one. A breakfast that releases its sugars gradually keeps your blood sugar

---

*While we often think of beans as humble foods—indeed, a packet of dried beans can feed a family for pennies—they are a nutrition powerhouse. They have no cholesterol and are virtually fat-free. They are rich in protein and have a surprising amount of calcium. They have cholesterol-lowering soluble fiber, just as in oatmeal, and plenty of omega-3 fatty acids, as in fish. They also have an extraordinarily stable release of their natural sugars for energy. The one thing beans don't yet have is a really good public relations firm to promote their many benefits.

more stable beyond lunch and well into the afternoon.[30] A dinner with a slow sugar release, say, lentil-barley soup, not only leads to a slow, steady sugar release in the evening, but also smooths out the absorption of sugars from your breakfast the next morning. If you swap the lentil-barley soup for instant mashed potatoes and white bread, these foods release their sugars quickly and disrupt your ability to handle sugars, both in the evening and the next morning.[31] So if you're inexplicably hungry after breakfast one day and just cannot wait for lunch, the reason may lie in your breakfast—or in the dinner you had the night before.

The explanation for the slow, steady sugar release some foods have goes beyond their fiber content. A more important factor is the shape of the carbohydrate molecule itself. If you could tease apart certain starchy foods, such as a bean or a pea, and examine them under a powerful microscope, you would find that their starch molecules are long and straight, and they stack up like a dense pile of wood. When you eat these foods, it takes time for digestive enzymes to work their way into these densely packed molecules, and they break them down to sugars slowly. Beans, peas, and lentils have plenty of this slowly digested starch. Quickly digested foods, however, are built from starch molecules that are branched, like a pile of small twigs. Enzymes easily find their way between the branches to break them apart.[24]

A REALITY CHECK

Not everyone needs to pay attention to the glycemic effect of foods. If you are slim and healthy despite a high glycemic index in the foods you eat, there is no reason to change. High-GI foods are not unhealthy in and of themselves, and there is no reason to avoid them if they are not causing you any problems.

Some popular writers have pushed ultra-high-protein diets on the theory that, since meat has no carbohydrate at all, it should not have any effect on insulin. Research has shown otherwise. Protein is a powerful stimulus for insulin release, just as sugar is. In fact, fish produces a bigger insulin release than popcorn. Beef and cheese cause a bigger insulin release than pasta.[32] A better answer is to chose high-fiber, natural plant foods.

Turn Off the Fat Genes

CHOOSING MEALTIMES TO SLOW SUGAR ABSORPTION

Chimpanzees do not gulp down eggs, bacon, toast, and a cup of coffee and then race to the office. They take their time and eat many times a day. The foods they eat are unprocessed and uncooked, of course, and are slow to release their sugars. When meals are small and frequent, blood sugar stays on an even keel.

If a meal is small and comes fairly soon after the previous meal, your body takes it in stride. There is no huge rise in blood sugar and no violent insulin response to try to cope with the sudden influx of energy. On the other hand, if you have a large meal after not having eaten for several hours, the sugar tends to rush into the bloodstream, causing a rush of insulin in response. It is easy for the body to overreact with too much insulin.

Years ago, it was observed that people who eat frequent small meals, instead of a few big meals, tend to be slimmer.[33] Of course, the important word is *small*.

FIGHTING THE FAT GENES

Whatever your genetic inheritance, you can choose foods for a better after-meal burn, and these same choices are likely to delay hunger's return. As you can see, our prescription is a far cry from the liquid diets, tiny frozen dinners, and diet sodas that are the main artillery many people use in the battle of the bulge. Instead, we will choose foods that work with our genes to promote easy and natural weight loss. Rather than skimp on food, we will shift the *type* of food we eat, favoring generous amounts of complex carbs and fiber, especially those with slow-release sugars. We'll be able to slim down while never going hungry. We'll give our overworked insulin a bit of time off while boosting our metabolism.

Chapter Six

How Genes Influence Your Exercise

I HAVE a friend named Brian who loves to exercise. It is his life. If I'm having dinner at an outdoor café, I'll see Brian running down the sidewalk and off into the distance. About the time I'm finishing dessert, he's running back the other way. Not only does he exercise an enormous amount, but he never looks tired. He is the Energizer Bunny in human form: he just keeps going and going and going.

As a youngster on the school track team, I learned that I am not the Energizer Bunny. Lacing up my shoes and setting off, I found running exhilarating at first. But after several minutes, it became more and more daunting. It took several minutes more for my reserves to kick in for a second wind. I never looked forward to running a marathon or anything remotely like it.

If you have known people like Brian, you have probably asked yourself how they can thrive on physical activity that would slay the rest of us. Well, recent scientific studies have made it clear that the difference between the Brians of the world and everyone else is, to a great extent, locked in their chromosomes. Endurance athletes get much of their stamina from their genes.

If you look at their muscles under a microscope, they actually look different from those of other people. They are rich in special muscle fibers, called Type I cells, which are endowed with a good blood supply—extra capillaries that bring in plenty of oxygen for energy. And that is these athletes' secret asset. As they run down the road, their blood vessels carry the oxygen they breathe straight to their muscles to power movement, which is why they tend not to tire easily.

These Type I muscle cells also have lots of lipoprotein lipase (LPL), the enzyme we met in chapter 4 that chisels fat out of particles in the bloodstream and passes it into cells for fuel. As a result, these athletes' muscles soak up fat and burn a tremendous number of calories during exercise. People with lots of Type I muscle cells are well suited to endurance sports, are slimmer than other people, and can eat more fat without putting on weight, all because their muscle cells are so eager to take in oxygen and fat from the bloodstream and turn them into energy.

The rest of us are different. If researchers take a sample of an average person's muscle tissue and examine it, they find mainly Type II cells, which have fewer capillaries, are less able to take on oxygen, and are more susceptible to fatigue.

The type of cell that predominates in your muscles—Type I or Type II—is not a matter of choice. It is, in large part, genetic. Not surprisingly, these genetic traits determine your athletic aptitude. Muscle samples from endurance runners are more than 60 percent Type I cells. Short-distance sprinters, on the other hand, have less than 30 percent Type I cells.[1] It was not the sport that changed their muscles. Rather, the type of muscles they were born with guided their choice of sport. No matter how long and how hard you train, a Type II muscle cell does not turn into a Type I, or vice versa, according to the best evidence science can muster.

DEFYING GENETICS, TURNING YOUR MUSCLES INTO ATHLETES

Let's say you did not get the muscles of a marathon runner and are not quite ready to race to Athens bearing news of a Greek victory. This is not a reason to join the Couch Potato Olympics. Believe it or not, you can change your muscle fibers to make up for what is missing in your chromosomes.

While exercise will not turn Type II cells to Type I, it will do the next best thing. If you start an exercise program and stick with it, it will increase the number of capillaries reaching each muscle cell by as much as 40 percent within a few months. The lazy Type II cells will become almost as vigorous as Type I's. In research parlance, Type IIb cells, which have a relatively poor blood supply, are turning into Type IIa cells, which are similar to Type I's.[1] Over time, your muscles will act very much as if you had the full genetic advantage.

The exercise I recommend is brisk walking for a half hour per day or one hour three times per week, increasing your time as you feel comfortable. The muscle type changes begin when exercise becomes rigorous and regular. Before we look at how to get started, let's take a look at some other benefits it will provide.

EXERCISE AND FAT-BURNING ENZYMES

In addition to reinvigorating your muscles, exercise has a wealth of health advantages. Yes, it burns calories. But that is by no means all there is to it. Exercise has other, arguably more important, effects.

Physical activity counteracts the fat-storage effect of LPL. As you'll recall from chapter 4, LPL extracts fat from the bloodstream and passes it either into fat tissue for storage—which means more body fat—or into muscle cells where it is burned for energy. Exercise does exactly what you might hope. It slows down the LPL enzyme in fat tissues, making it harder to store fat, and increases LPL's activity in muscles, pushing fat into muscle cells to be burned.

The effect can be dramatic and quick. Researchers at Cedars-Sinai Medical Center in Los Angeles measured LPL activity in a group of athletes by removing samples of fat and muscle with a tiny needle. Their muscle LPL was twice as active as their fat-tissue LPL, meaning these enzymes were aggressively pushing fat into muscle cells to be burned for energy. The researchers then asked the athletes to *stop* exercising. Their LPL got lazy in a hurry: after they had sat around for just two weeks, the amount of LPL in their fat tissue increased to more than four times that in muscle. Inactivity had forced their bodies into fat-storing mode.[2]

The good news is, you can get LPL back into gear just as quickly. A single bout of exercise causes LPL activity in muscle cells to spike up to ten times its previous level.[2] A group of researchers asked a group of volunteers to exercise one leg and not the other, using a machine for knee extensions. In the exercising leg, LPL activity increased 62 percent compared to the unexercised leg.[3] The effect is temporary, but regular physical activity keeps it going strong.

GET YOUR CALORIE BURN

Here are some activities with the number of calories they burn for a 150-pound adult. For a 100-pound person, subtract one-third. For a 200-pound person, add one-third.

A brisk half-hour walk	220
A leisurely half-hour bicycle ride	120
A half-hour bicycle ride at moderate speed	200
A half-hour leisurely swim	140
A half-hour fast swim	250
A slow half-hour jog	370
A quick half-hour jog	460
Running in place for half an hour	325
A half hour of singles tennis	200
A half hour of cross-country skiing	350
Jumping rope for half an hour	375

THE METABOLISM-BOOSTING EFFECT

Exercise can even help make up for the low metabolism your genes may have given you. Researchers have tested the effects of a long bout of aerobic exercise, finding that it pushes your metabolism up—and it stays high for at least a day after the exercise. Long-distance runners show how far this can go. They have roughly a 10 percent higher resting metabolic rate (RMR) around the clock, compared to sedentary people, which keeps them burning calories quickly and easily.[4]

After a workout, your body strengthens your tired muscles and repairs their stresses and strains, like mechanics working furiously to repair a car after a race. Building new proteins burns a lot of calories and can account for 10 to 15 percent of your RMR.[4] Exercise protects you from the gradual fall in metabolism that happens to many people as the years go by.

Not surprisingly, how much you benefit from exercise depends on your genes. Researchers in Quebec, Canada, studied groups of identical male twins, measuring how many calories they burned on an exercise bicycle, which was checked by measuring how much oxygen they consumed minute by minute. If one man burned calories quickly during exercise, his twin was much the same. If one burned calories more slowly, his twin was again similar.[5]

In a longer-term study with the same group, using a vigorous, three-month exercise program, each individual again responded much like his twin. If one man trimmed away five pounds of fat, his twin lost nearly the same amount. If one lost ten pounds, his twin had a similar benefit. Another twin pair might show a greater or lesser effect, and genes were apparently the reason.[6] Here is the good news: everyone in the study lost weight. Some lost it faster, but exercise benefited everyone.

REGULATING APPETITE

Exercise also helps control your appetite. While you might assume that intense physical activity would be followed by a voracious appetite, the opposite is generally true. Exercise takes your body out of the mode of eating and digesting and makes you unlikely to binge.

Some years ago, I had a dinner scheduled with some friends from the hospital staff. Before dinner, several of us decided to go to a racket club for some vigorous racquetball and swimming. When we got to the dinner, none of us ate much. Far from "working up an appetite," exercise does exactly the opposite. Vigorous physical activity triggers a response in the body similar to the "fight or flight" reaction, in which the body is set for action, and appetite and digestion are essentially turned off. The effect lasts only hours, but it is very noticeable.

If you find yourself making late-night refrigerator runs, schedule your exercise routine to head this off. A vigorous walk or run will also help you sleep better, so you will be better able to resist temptations the next day.

See chapter 3, and be sure to speak with your doctor to be sure exercise is safe for you.

HOW TO BEGIN?

So how do we get started? Let me offer a few suggestions:

First, see your doctor, and start slowly. It is easy to jump into vigorous exercise too quickly, and if your heart, lungs, and joints are not yet up to it, you can cause yourself serious injury, or worse. So if you are overweight, have any medical condition, or are over forty, let your doctor advise you on how to begin an exercise program.

Also, if you have been on a low-calorie diet, it has probably slowed your metabolism. Although most people get a metabolism boost from exercise, it can have the opposite effect if you've been dieting. Put off any vigorous physical activity for now. Get your meals in shape for a few weeks with no artificial calorie limit, then start your exercise program once your metabolism is back to normal.

Here are the elements for success:

1. Start easy. Begin with a brisk half-hour walk per day, or as much exercise as your doctor will allow. If a daily walk is not convenient, go three times a week for an hour. This sounds simple, but it is more than enough to do the job.

   Set aside the "no pain, no gain" mentality. Increase your activity gradually and comfortably. The point is to enjoy the use of your body so you will want to keep it up. Pick a place to walk that is enjoyable for you. Enjoy the sights, sounds, and smells, and feel free to substitute any equivalent activity.

2. Make it fun. Many adults view exercise as self-imposed torture. They tend to pick tedious or even painful exercises, and they often do them in solitude—a perfect recipe for failure. Yet not so many years earlier, when they were children, they couldn't wait to run around the neighborhood or join in sports and games. Fun is what helps you stick with it. If you like dancing, gardening, bike riding, a run with your dog, or a vigorous walk in the woods, then that is the exercise that's right for you.

Turn Off the Fat Genes

3. Be with other people. If you regularly walk or work out with a family member or friend, you will find it more enjoyable and easier to continue than if you are by yourself. Making it a scheduled social event decreases the possibility of your drifting back into sedentary living. Aerobics classes at a local health club are a wonderful way to get a great workout and to pick up your pace as your fitness increases, while building new friendships that can take the pain out of exercise. If you get to know an instructor or two, they will help keep you on track.

4. Be judicious about home equipment. Many people have hundreds of dollars' worth of equipment gathering dust in their basements when what would be more helpful is a decisive change in diet along with a regular walking program.

   This is not to say there is no value in equipment. The *Journal of the American Medical Association* recently published a study in which people were asked to exercise for forty minutes daily in either one continuous bout or in ten-minute sessions four times a day. The people doing the shorter sessions lost weight just as well as those who did the single long exercise bout, and what made it practical was having a treadmill at home. No one would go out running four times a day, but hopping on and off a treadmill at home for a quick exercise session a few times a day was easy and fun.[7]

THE LIMITS OF EXERCISE

To be effective, exercise must be used *with* changes in your diet, not in place of them. Alone, it is not particularly useful. A single pound of body fat stores, believe it or not, more than 4,000 calories, and it would take hours of exercise to burn it off. A study of weight-loss programs showed that those using exercise alone led to an average weight loss of only six pounds in sixteen weeks, while those using diet alone caused an average weight loss of about twenty-four pounds in the same time. Using diet and exercise together helps you maximize your results and makes it easier to keep the weight off over the longer term.[8]

Most important, don't count on exercise to counteract the effects of a fatty diet. A friend of mine always does an extra workout in advance of holiday parties, in anticipation of overeating to come. But a single bout of eating brings in far more calories than a workout can burn. A brisk half-hour walk burns only 200 calories or so, and you can easily take in that many calories in five or ten minutes of eating.

Seagulls provide a wonderful demonstration of how easily eating can overcome the calorie burn of exercise. Along the Atlantic coastline, gulls find clams, pick them up in their beaks, climb to a height of thirty feet or so, and drop the clams onto the ground to break them. It is a model of inefficiency. It may take a long time to find a clam, and perhaps a couple of tries to break the shell, with a prolonged struggle to keep the prey away from competitors. Yet the calorie expenditure of all of this labor is apparently more than compensated for by the rather modest calorie content of the clam. Similarly, an athlete can be winded from a serious workout, but even a normal-sized meal can easily replace all the burned calories. The moral of the story is that eating brings in far more calories than exercise burns.

Some people dangerously imagine that exercise can somehow work off blockages growing in their arteries resulting from eating meats, cheese, eggs, and other fatty, cholesterol-laden foods. This strategy does not work at all. In the Korean conflict, autopsies showed atherosclerosis in 77 percent of American GIs, despite their physical fitness, regular exercise, and young age (averaging about twenty-three years), while their Korean counterparts had much healthier arteries. The difference, of course, was diet. American diets are rich in meats and dairy products, while Asian staples are rice, noodles, and vegetables, with much less food from animal sources. Exercise is an important *part* of programs to reverse atherosclerosis, but it is not much use without a major change in diet.

Some years ago, Bill Clinton set out to trim his waistline with an exercise program. He strapped on his athletic shoes and set off down the street. As he jogged along, he spotted a McDonald's and stopped in for a burger before jogging back to the White House. Not surprisingly, the exercise program didn't work well. A few bites of a burger provide more calories than you can burn off in a half hour of exercise. To his credit, President Clinton later asked the White House chef to stock vegan Boca Burgers, which gave his weight-loss program a badly needed shot in the arm.

EXERCISE AND NUTRITION

A word to athletes: Your body needs good nutrition, not only to power your extra activity but also to repair and rebuild after your workout.

As we saw in chapter 5, endurance depends on your muscles having plenty of glycogen, which is built from the natural sugars released from carbohydrates. Endurance athletes use special "carbohydrate loading" methods to pack glycogen into their muscles. In the original method, athletes first emptied out their glycogen stores by exercising to exhaustion and then eating a low-carbohydrate diet for three days. Then for another three days, just before competition, they ate a high-carbohydrate diet to pack in glycogen. Although it works, it is a tough regimen, and it is hard to train on the low-carb days. An equally effective but more acceptable method is for athletes to taper their training during the week prior to competition, and to greatly boost their carb intake during the final three days before an event.[9]

Numerous studies have shown that high-carb diets really do improve endurance, and a high-carb, low-fat meal eaten about three to four hours before a competition boosts performance. Much the same applies after an athletic workout, when your body is replenishing its glycogen stores. This is one time when high-glycemic-index foods, such as white bread, potatoes, and sweetened drinks, actually have demonstrable value, reducing muscle breakdown and cutting recovery times.[9]

You will also need more protein, which is found in a surprisingly wide array of healthful foods. Most green vegetables, such as asparagus or broccoli, are about 40 percent protein, as a percentage of calories. The hitch, however, is that vegetables are not at all nutrient dense, so you would have to eat generous servings of vegetables to boost your protein intake substantially.

Legumes (beans, peas, and lentils) supply a significant amount of protein and are more nutrient dense. Most varieties yield about fifteen grams of protein per cup. Soybean products push protein higher and are available in fat-free versions, such as textured vegetable protein—defatted soy nuggets sold in all health food stores, which substitute for ground beef in chili, tacos, or spaghetti sauce. You will also find soy-based protein supplements at health food stores. Most contain about twenty-five grams of protein per one-ounce serving, with virtually no fat.

How Genes Influence Your Exercise   103

Pasta, believe it or not, contains a fair amount of protein. Some varieties have about ten grams of protein in every two ounces of dry pasta. Be sure to steer clear of fatty toppings.

Meats, poultry, fish, eggs, and other animal products contain protein, but typically bring plenty of fat along with it and are devoid of fiber, making them inferior choices. Check the chart on page 132 for a listing of healthful high-protein foods.

TURNING YOUR GENES ON AND OFF

Modern life makes many of us sedentary. Our grandparents walked; we drive. And televisions, computers, and desk jobs give our bodies little to do. Of course, you do not necessarily have to exercise to lose weight. In our research studies, we have seen quite profound weight loss in individuals who change their diets, whether they exercise or not. But, every movement you make, whether it is blinking your eyes or lifting a grand piano, burns some calories. The more you move, the more calories you burn. Exercise increases your muscles' blood supply and gives them endurance, increases their LPL enzymes to help you eliminate fat, and boosts your metabolism so that calories are burned more quickly, not only while you are exercising, but also afterward for a time. Exercise regulates your appetite and preserves your muscles. Muscle tissue is much better than fat tissue at burning off calories, so if you keep your muscles from wasting away, your body will be able to burn calories faster.

A final word comes from researchers at the University of California at San Diego who examined how body weight runs in families. They made an encouraging finding: however strong the inheritance for weight problems might be in the people they studied, exercise helped counteract it.[10] So even if everyone on your family tree is struggling with extra weight, a combination of the nutrition recommendations in this book and regular exercise can help you carve out your own healthy body.

Part Two

Manipulating Gene Action

Chapter Seven

The Three-Week Diet Makeover

T HE MOST important step in orchestrating gene effects is changing the nutrients your genes have to work with. Usually, this means making some major improvements in your diet, and this will be easier for some people than for others. In our research studies, we have found that, while some people take to diet changes easily, others need a bit of hand-holding.

In this chapter, let me share with you the techniques we use to help people start new habits, break old ones, and build the foundation for long-term success, making the transition fun and engaging. The two fundamental "tricks" we use are, first, making major diet changes rather than minor ones, and second, focusing on the short term.

First, making major changes: We don't put our toe into the swimming pool. We jump in. We change our diet dramatically so we can see the quickest and greatest possible results. If we were to simply dabble with minor adjustments to the diet, the results would be too modest to reward our effort.

Second, we focus on the short term. It is easier to make a short-term commitment than a long-term one, so if we limit our dietary experiment to three weeks or so, we can commit to it fully. At the end of three weeks, see how you feel. If you like the way things are going, you can continue. If not, you can stop.

People quitting smoking often discover the value of these principles. If they go cold turkey, they may find the first week or two a bit difficult, but it soon becomes much easier, and after a month or so, you couldn't pay them to start smoking again. If, however, smokers simply cut down, teasing themselves every day with the very taste they are trying to get away from, they are constantly at risk of backsliding. Also, people quitting smoking do not have to plan whether they will still be nonsmokers twenty years down the road. All they need are a clean break and a focus on getting through the next couple of weeks. Diet changes are the same way.

Incidentally, it is much easier to break a food habit than to quit smoking, as I learned from having done both. While smokers do not find many entirely satisfactory substitutes for tobacco, innumerable healthful foods can substitute for unhealthy ones, as you will see in the recipe section.

REEDUCATING YOUR TASTE BUDS

An additional reason for changing the diet dramatically is that, if you have a weakness for any kind of food—sweet, fatty, salty, or whatever—a major diet change will reeducate the taste buds. Your preferences are mainly set, believe it or not, by what you have been eating during the preceding three weeks or so.

One of the most surprising discoveries in the science of appetite is that *tastes require maintenance.* If you do not regularly apply sweets or salty foods or fatty foods to your tongue, they lose some of their appeal. Just as a person's desire for a cup of coffee drops dramatically if the coffee routine has been broken for a month or so, the sweet, salty, and fatty tastes are maintained by your current habits. If you don't feed them, they die.

If you eat salty foods, for example, your taste buds *expect* a salty taste and will be unsatisfied without it. If you have been avoiding salt for whatever reason, your taste buds come to prefer a lower-salt taste. It depends entirely on what your taste buds come in contact with. It does not work to use much less salt in cooking and then salt up the surface of the food at the table. The bit of salt touching your taste buds reawakens the salt taste. Although taste-changing is a two-way street, setting your tastes to prefer less salt is a bit slower than increasing your salt taste.[1] But both are easy to do.

Turn Off the Fat Genes

Sweets work the same way. When researchers prepare beverages with less and less sugar, they find that people quickly develop a taste for the milder flavor.[1] If you don't maintain the sweet taste, it diminishes. It may never be gone, but it does quiet down.

Ditto for fats. When people eat foods that are lower in fat, they soon come to like the lighter taste, as you know if you have ever switched from whole milk to 2 percent milk or skim. At first, skim milk seems watery. But after a week or two, it tastes perfectly normal. If you then taste whole milk again, it seems too thick, almost like paint. This is not to say that skim milk is a healthful food. Indeed, most of its calories come from nothing but lactose sugar, and dairy products—skim or otherwise—are linked to various health problems. But this common experience illustrates how quickly and dramatically our tastes can change. Two or three weeks is all it takes.

Researchers in Philadelphia found that people can reduce their preference for fatty foods by simply eliminating fats that are added to the surface of foods, such as salad dressing, butter, margarine, or mayonnaise. The less often these foods touch your taste buds, the less you care about them. Your preferences will continue to favor lower-fat foods unless you let the higher-fat varieties seduce them again.[2]

A couple of notes of caution: First, seductions are everywhere. Even the occasional fatty meal can push up your fat taste. Second, artificial fats do not help retrain your tastes. Reduced-fat margarine, olestra, "low-fat" cheeses, etc., seem like the real thing to your taste buds and actually encourage fatty tastes.

A DIET MAKEOVER

Chances are, your goal is not simply to break a chocolate craving or to cut down on sweets. You are aiming for major success in losing weight. The same principles apply.

Not long ago, we did a research study comparing different nutritional approaches to diabetes, in cooperation with Georgetown University Medical Center.[3] We aimed to make sweeping diet changes to get our volunteers into better shape as rapidly as possible, but we focused on the short term, with group meetings weekly.

CHANGING THE TASTE FOR FAT

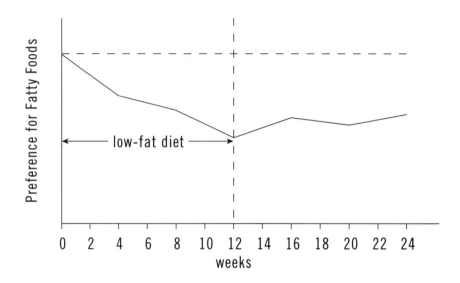

Twelve weeks on a low-fat diet that also omits added fats (e.g., salad dressings, table spreads, mayonnaise) causes a marked reduction in the taste for fatty foods, which persists even when individuals are free to return to their normal diets.

SOURCE: Mattes RD, "Fat preference and adherence to a reduced-fat diet," *Am J Clin Nutr* 57 (1993), American Society for Clinical Nutrition: 373–81.

The experimental diet was quite a switch from the menus most people are used to. It eliminated animal products and kept vegetable oils low, while allowing essentially unlimited amounts of whole grains, pasta, bean dishes, vegetables, and fruits. These changes were designed to repair the participants' insulin sensitivity as quickly as possible.

In the first week or two, some people liked the diet very much, while others had problems—some recipes didn't seem to turn out right, it was harder to eat at restaurants, their spouses did not know what to do with the new foods in the refrigerator, etc.

Turn Off the Fat Genes

But after about two weeks, the complaints stopped. All had figured out which recipes they liked and found restaurants with good menu choices. Their spouses came to enjoy the foods, too. By twelve weeks, the challenges of breaking old habits had melted away. The average participant had lost sixteen pounds, most were able to cut back or even discontinue their medicines, and they were having fun exploring new tastes, new foods, and new restaurants. When the study was finished, many were quite sad to see the dietary adventure end. They wanted to keep going and see how far their weight loss, health improvements, and new taste explorations could go.

We used a similar regimen to study the effect of diet on menstrual pain and premenstrual symptoms (for many women it is remarkably effective).[4] We found that about one-third of our research participants had no trouble at all with the diet change. After all, they had already been on more taxing weight-loss diets—starvation diets, grapefruit diets, cabbage-soup diets, diet shakes, and so on—which made our regimen seem surprisingly easy. The other two-thirds, however, did grumble a bit. Just as in the diabetes study, giving up some familiar foods—even temporarily—and trying new tastes put them on unfamiliar ground. "Is a life without cheese or the occasional bit of chicken worth living?" they asked. After a week or two, however, most started to feel better. Their energy level improved. Old aches and pains melted away. After the first month, many noticed that their menstrual pains were milder or nonexistent. And they lost weight—about a pound a week. The more weight they had to lose, the more quickly and dramatically it came off. They appreciated how good the new diet made them feel, and their tastes changed.

THE THREE-WEEK RUT-BUSTER

Whether you aim to turn down a specific craving or make a major diet change, the key is not to use daunting words like *forever* or *permanently*. Instead, we aim to keep our taste buds clear of whatever foods we are concerned about for some three weeks. This is not a long-term commitment. All we are doing is a brief taste-retraining exercise.

Before we start, let's make sure that what we are doing is worthwhile. There is no point in giving up chocolate, for example, if you are continuing to eat other greasy foods—chicken wings, cheese, salmon, or pork chops, all of which have so much fat that avoiding chocolate is a complete waste of time.

Assuming, however, that you do have a sensible reason for breaking out of a taste rut, remember our two keys: give it 100 percent, and focus on the short term.

- Select a three-week period, and during this time keep your taste buds as clear as humanly possible of the foods you're trying to avoid. If they are camping in your cupboards or refrigerator, eliminate temptation by giving or throwing them away.

  If you have no particularly troubling cravings or food habits, but simply want to get going with the best foods for permanent weight control, you will want to use the diet changes described in chapter 8. For a three-week period you will build your diet from generous amounts of whole grains, vegetables, fruits, and legumes, while eliminating animal products and added oils. If you are looking to jump-start your weight-loss success, a three-week commitment to these foods will do you a world of good, prying your taste buds away from greasy tastes, dramatically cutting your fat intake, and improving your insulin sensitivity. You will find plenty of food examples in the recipe section.

- Take advantage of social support. In 1995, I published in the *Archives of Family Medicine* the results of a study on why some researchers succeeded in helping individuals change their diets, while others failed.[5] Social support was often the key. Friends and family keep us afloat during times of uncertainty. If they give you a hard time about the diet change you are making, the process is much more difficult. But if you can bring them along as you change your diet, they are likely to benefit, just as you will. After all, the dietary changes described in this book are not only good for people seeking to shed some weight. They are also great for lowering cholesterol levels, preventing cancer, and all-around better health. So don't be shy about ask-

ing your spouse, partner, or family to join you in experimenting with some new foods.

You can also get some helpful social support by attending
cooking classes or by joining a local vegetarian society, where you
will learn about new tastes and meet people who are committed
to healthy eating habits.

- Jump-start your diet change. Researchers sometimes start by
  sequestering their research participants in a hotel or hospital
  ward to quickly teach them new tastes while removing all
  temptations. While that may not be practical for you, you can
  do the next best thing, which is to test out some recipes, food
  products, and restaurants well before your three-week diet
  makeover begins. It helps to plan out which foods you like and
  which groceries to have on hand.

- Be sure to get plenty of rest. You have no doubt noticed that
  when you are not well rested, you tend to lose any restraint
  you may have for food. You'll have snacks that might have
  been off limits at other times. This is not simply a reaction to
  stress. A lack of sleep actually decreases your taste sensitivity,
  meaning that you'll feel less satisfied with normal flavors and
  normal amounts of food and will seek out stronger flavors to
  feel satisfied.[6]

- Be on the lookout for things that push you back into old
  habits. Does alcohol destroy your resolve? Do spicy meals
  make you crave sweets? Do certain friends or certain restaurants lead you into a grease-fest? If so, avoid them for now.

There is one thing this exercise will not do: it will not erase memories
of tastes you once loved. They will be much less compelling, but they will
not be gone. Just as a smoker who is delighted to have quit still remem-
bers the taste of a cigarette, the same is true of those who have broken a
food "addiction." You will remain vulnerable to slipping back into old
habits. Should that happen, don't berate yourself. Simply do another
three-week rut-buster when you are ready, and you will be back on track.

This technique applies only to changing your tastes. If you find yourself eating compulsively, let me refer you to chapter 3, where we dealt not with tastes but with appetites.

LEAVING A RUT, FINDING A GROOVE

In the next chapter, we'll take a closer look at the best foods to focus on in our diet makeover, and we'll be well on our way to success.

Chapter Eight

Food Choices for Optimal Weight Control

THE PHYSICIANS Committee for Responsible Medicine has studied how foods affect health in many different ways. We have examined which diet patterns are most powerful for cutting cholesterol levels, getting hormones into balance, managing diabetes, and dealing with many other problems. Over the long run, a change in diet often rivals the power of drugs or surgery.

Emerging from our research is a weight-control method that counters fat-building genes and bolsters "thin" genes. It works without any artificial restrictions on portion sizes—and even without exercise, although exercise adds to its benefits. Rather than focusing on the amount of food you eat, we focus on the type. Because it is based on a fundamental menu change, its effects can essentially be permanent.

The diet changes we use are much easier to follow than calorie-restricted diets, fad diets, or ultra-high-protein diets. Our research participants learn new tastes and set aside old habits, and before long, new and healthier ways of eating become second nature.

GENES AND WEIGHT-LOSS SUCCESS
Genes have undoubtedly affected your weight-loss efforts so far, influencing which foods you gravitate toward, how actively your body stores fat,

and how quickly you can burn it off. One person's taste may call for chocolate, while another loves vegetables. And while one person may find that even tiny deviations from an optimal diet cause seemingly instant weight gain, another has more latitude.

But the genetic differences that make us individuals do not change the broad guidelines for a healthful diet. No one genetic type requires ice cream while another thrives on double bacon cheeseburgers, any more than anyone needs cigarettes or Scotch, despite genetic influences on how attractive or dangerous these substances might be. In this chapter, we will set as perfect a nutritional prescription as possible and look at how our diet goals interact with our genetic assets and liabilities.

In the preceding chapter, we looked at how to jump-start a diet change. If you have not thoroughly reviewed that chapter, please do so before you begin this one. The emphasis on a short-term but committed focus leads directly to the principles in this chapter.

Let's build a menu based on the key genes we've looked at in chapters 1 through 6. Our goals are to choose foods that

- fit our taste preferences and vulnerabilities
- bring weight loss that is neither too fast nor too slow, so our leptin system works normally
- fool chromosome 8 by depriving your fat-storage (LPL) enzymes of storable fat
- prevent insulin excesses that could interfere with fat burning

RESPECT YOUR TASTE TYPE

In any sensible weight-control program, you will want to greatly increase your intake of vegetables and fruits, as they are low in fat and calories, loaded with fiber, and rich in nutrients. But you will need to respect your taste type.

An old doctor friend of mine always insisted that he hated vegetables. He could not abide broccoli or cauliflower, and spinach held absolutely no attraction. His meals were heavy with roast beef and poultry, and his side dishes were mainly potatoes or bread slathered with butter. His lifelong weight problem had seemed insoluble. As you might have guessed, genetically he was a PROP supertaster, acutely sensitive to the bitter parts of vegetables, which is why they never figured in his menu. My advice to him was to stock his pantry with plenty of fruits, whole grains, and bean dishes, and when it came to vegetables, to favor carrots, sweet potatoes, and any others that would not trigger his bitter sensors. Several of the recipes in this volume are ones he enjoyed. If he got up the courage to steam some spinach or other greens, I suggested he sprinkle on a bit of lemon juice, which, for some reason, eliminates their bitter flavor. After some experimenting, he found that some healthful foods tasted pretty good after all, and he managed to lose quite a bit of weight.

The moral of the story is to avoid forcing yourself to eat vegetables you are not genetically programmed for. This is especially important for children, whose genetic tendencies are clear and decisive, driving them straight toward sweets and away from anything with even a hint of bitter flavor.

If you find yourself at the opposite extreme—where absolutely everything tastes great, and more of it, please—you may be a PROP nontaster. The danger you face is that your somewhat insensitive taste buds, in their search for satisfaction, may lead you to overeat. In this case, you will be fine if you stock your pantry with whole grains, vegetables, fruits, and legumes (beans, peas, and lentils), because these foods are not at all calorie dense, so an occasional overindulgence causes no problem. Spaghetti with a light tomato sauce, veggie chili, and healthful soups, such as split pea, lentil, black bean, or minestrone, are so low in fat that overdoing it a bit poses no risk. However, you will run into trouble if you stray into fatty or sugary foods. Cheese, meats, potato chips, and other calorie-dense foods will easily contribute to body fat.

In a nutshell, the advice is simple: if the PROP-taster gene keeps you from tolerating broccoli, spinach, grapefruit, and other bitter foods, plenty of sweeter choices are in the produce aisle, from carrots and sweet potatoes to oranges, apples, and pears.

If, on the other hand, almost everything tastes good to you and your problem is limiting quantities, keep well stocked with whole grains, beans, vegetables, and fruits, as these foods are not calorie dense and tend not to promote overweight. If you want to change your tastes, see chapter 7.

Food Choices for Optimal Weight Control   117

SPARING LEPTIN: PACING YOUR WEIGHT LOSS

A low-calorie diet will rob you of leptin, encourage an out-of-control appetite, and bring a dispiriting weight gain soon after your diet is over. An optimal weight-loss speed—neither too fast nor too slow—keeps your leptin system in gear and makes it easier to maintain your progress and avoid backsliding later on. Avoid a severe calorie restriction, no matter how motivated you may be to lose weight.

Very low calorie diets not only slow down your calorie-burning speed, but they are also linked to gallstones, pebblelike crystals that form in the gallbladder, a small pouch below the liver. Anyone who has extra weight has a higher than average risk for gallstones, and during times of rapid weight loss—over about three pounds per week—they become particularly common. About one in four people on very low calorie diets develops stones, and weight cycling seems to increase risk.[1-4] Although stones often cause no symptoms, sometimes they are painful and require stone-dissolving medications or surgery.

The trick to controlling your weight-loss speed is to focus on the *type* of food you eat, rather than the amount. There should be no artificial calorie limits or tiny portion sizes. We are not aiming to starve weight off but to ease it away by restoring your body's natural metabolic balance.

BUT REALLY, HOW MUCH SHOULD I EAT?

Many people look for a calorie limit to help them in meal planning. However, the number of calories you need depends on how many you burn in a day, and that depends on your size and activity. The larger and more active you are, the more calories you burn. Rather than follow a specific calorie limit, I suggest you let your natural hunger drive guide you. Your hunger sense is much more reliable once your diet is tuned up, as we will see.

If you really feel lost without a specific number of calories to shoot for, use the Rule of 10. Simply take your ideal body weight in pounds and multiply it by 10. This gives you the minimum number of calories you should consume in a day. For example, if your ideal body weight is 150 pounds, you should take in at least 1,500 calories each day. Any less and you will be headed for a slowed metabolism and a tendency to binge.

Remember, the ten-calorie rule is a *minimum*. You can, and almost certainly should, take in more calories than this, and you will still lose weight.

IDEAL WEIGHT LOSS: NOT TOO FAST, NOT TOO SLOW

The most healthful weight loss occurs at around a pound per week. To do this, avoid any strict calorie limit. Focus on *what* you eat, rather than how much.

If you feel you need a specific guideline for your calorie intake, set a *minimum* of ten calories per pound of body weight per day. This will prevent the problems dieters face when they eat too little.

FOODS THAT BLOCK YOUR FAT-STORAGE GENES

As we saw in chapter 4, body fat starts out as fat in the foods you eat. Lipoprotein lipase (LPL) enzymes, built by chromosome 8, pluck fat from your bloodstream, so it can be packed into your fat cells.

Although this enzyme design is hardwired into your chromosomes, you can bring your fat-accumulating machinery nearly to a grinding halt by taking most of the fat out of your diet. Since your LPL only works when it sees particles of fat going by, if you build your menu from foods that have little fat in them, your LPL can just about take the day off.

Your best bets are foods that satisfy your appetite but have no more than traces of natural fats. Four broad categories of foods have these characteristics:

- *The Whole Grain Group,* such as rice, noodles, bread, corn, cereals, oatmeal, barley soups, and others are extremely low in fat, but loaded with fiber. While many people are fearful of the carbohydrate content of these foods, they are the best friends of anyone seeking to lose weight. The fat content of a bowl of brown rice is one-fourth that of skinless chicken breast, and its fiber content is far higher. Below we will see how to select among them wisely.

  Years ago, Dr. Walter Kempner set up a weight-loss center at

Food Choices for Optimal Weight Control   119

Duke University using a diet containing enormous quantities of rice, fruit, and virtually nothing else. It came to be known as the rice diet, and it caused rapid and substantial weight loss. It also led to a new appreciation of other simple foods. After seemingly endless bowls of rice, the taste of an added tomato or green leafy vegetable is welcome indeed. Here, we will not attempt to use grains as your only weight-loss tool. But they are indeed valuable and powerful.

- *The Legume Group,* composed of beans, peas, and lentils, brings in plenty of protein and other nutrients, but most varieties are only about 4 percent fat, as a percentage of calories. Whether they flavor a soup or make a whole dish, such as baked beans, they give your body plenty of nutrition, but leave your fat-storing LPL little fat to work with.

  These foods also work as excellent replacements for meats. Switching a meat chili to a bean chili, for example, can save you a huge load of fat that would otherwise end up on your thighs.

- *The Vegetable Group* is nutrient rich, with plenty of protein, vitamins, minerals, and fiber, but little fat. Broccoli, carrots, spinach, sweet potatoes, and virtually all other vegetables can supply the body with the healthful fat traces it needs, without the excesses in many other foods.

- *The Fruit Group:* Apples, bananas, cherries, mangoes, oranges, papayas, peaches, pears, and nearly any others are extremely low in fat, and rich in vitamins and fiber.

Collectively, these are called the New Four Food Groups. They are low in fat, with no cholesterol at all, and they are rich in healthy, slow-release carbohydrates: all the ingredients for a healthy weight-control menu. There is plenty of protein in grains, beans, and vegetables, and calcium in green leafy vegetables and legumes. The one nutrient that requires a bit of planning, vitamin B_{12}, is easily found in fortified products (e.g., breakfast cereals, soy milk, meat analogs) or any common multivitamin (more on ensuring complete nutrition in the next chapter).

The name New Four Food Groups was coined in 1991, when the

groups were first proposed as a replacement for the 1956 "Basic Four" food groups promoted by the U.S. Department of Agriculture, and they are a much healthier dietary plan than the Food Guide Pyramid, which has been used by the government since 1992.

STEERING CLEAR OF MEATS AND OTHER FATTY FOODS

To keep fat-storing LPL as unemployed as possible, it is best to avoid animal products and added oils. If you avoid animal products, you will eliminate 100 percent of the animal fat from your diet. This is an extremely powerful choice. You may have read about Dr. Dean Ornish's use of vegetarian diets to reverse heart disease, which revolutionized cardiology in the 1990s.[5] When he eliminated animal products in an overall healthy lifestyle, he found that coronary arteries actually *reopened* on their own, without surgery. The powerful diet and lifestyle program also caused tremendous weight loss. One research participant lost nearly one hundred pounds during the first year, and the average weight loss—including those who had a lot of weight to lose and those who were near their ideal weight to start with—was twenty-two pounds in a year.

The key to this success was to avoid even supposedly "lean" meats—skinless chicken breast and all the rest. These products may seem close to fat-free, but as you can see in the table on page 122, they are anything but, and most other cuts of poultry, fish, or beef are even worse. None of these products lowers cholesterol levels effectively or is of much use for long-term weight control.[6]

In our research studies, we have found that it is much easier to exclude fat when you don't have meats or other animal products to contend with, and a pure vegetarian (vegan) diet, used over a twelve-week period, causes about double the weight loss of a low-fat, nonvegetarian diet.[7]

Most foods from plants are naturally low in fat, but there are a few exceptions: avocados, olives, nuts, seeds, and some soy products pack a fair amount of fat. Keep these to a minimum. The concentrated oils used in frying, in some salad dressings, or in baked goods are even worse, insinuating their way into your body as easily as the fat in Spam or ice cream.

You will never eliminate vegetable oils completely. There are always traces of natural oils in grains, beans, vegetables, and fruits, which supply

FAT IN ANIMAL PRODUCTS VERSUS PLANT FOODS

(as a percentage of calories)

Beef top round, lean	25%	Baked beans	4%
Cheddar cheese	74	Cauliflower	6
Chicken breast, skinless	23	Lentils	3
Chicken egg	61	Potato	1
Chinook salmon	52	Rice, white	2
Halibut	19	Spaghetti noodles	4
Pork tenderloin, lean	26	Spinach	9
Turkey breast, skinless	18	Sweet potato	1

SOURCE: Pennington JAT, *Bowes and Church's Food Values of Portions Commonly Used,* 16th ed. (Philadelphia: Lippincott-Raven, 1998).

the tiny amount of fat—about 3 to 4 percent of your calories—that your body needs.

Your genes will determine how much oil you can get away with. Some people eat generous amounts of guacamole, olive oil, and salad oils and still manage to avoid obesity, at least for a while. Others need to be much more careful.

IS OLIVE OIL AS BAD AS CHICKEN FAT?

Animal fat is every bit as unhealthy as scientists have said. It clogs your arteries and expands your waistline. But vegetable oils can fill your fat cells, too. Over the centuries, people have exercised their creativity in finding ways to concentrate vegetable oils into forms nature never envisioned. An ordinary olive has less than a half gram of fat. However, if you take a few hundred olives, throw away the pulp and fiber, and concentrate the oil, you can fill a bottle with it and fill out your waistline as well. Vegetable oils are much healthier for your cholesterol level and your heart, but for optimal weight loss, you will want to minimize vegetable oils, too.

For now, the best advice is to keep all fats to a bare minimum. Even though natural plant oils are much safer than animal fats, I encourage everyone to be cautious about them, as over the long run they contribute to weight problems. If you set aside animal products and keep vegetable oils to a bare minimum, your LPL has a lot less to work with.

Shortly, we will look at how the New Four Food Groups fit into actual meals. First, let us bring in the final basic principle for planning our meals.

CHOOSE FOODS THAT RELEASE SUGARS GRADUALLY

Most people with weight problems lose weight by simply building their menus from whole grains, legumes, vegetables, and fruits. Their bodies let fat go because they are not letting LPL store much of any fat, and their calorie intakes have probably dropped a bit, too, because these foods are naturally low in calories.

However, some people need to go one step further and look at how quickly their meals are releasing natural sugars. Foods that give you a slow, steady sugar release give you a healthy after-meal burn—the thermic effect of food—without an overly high insulin release that could block your fat-burning. Here are your best choices:

- For grains, the best choices are those with their natural fiber left intact—brown rice, as opposed to white rice, or whole-grain bread instead of white bread, for example. Oddly enough, pasta has a slow sugar release, even if it is ordinary white pasta. The reason, apparently, is in the process that packs white flour into noodle shapes.
- Virtually all legumes (beans, peas, and lentils) are excellent choices, with a slow, steady release of energy.
- The same is true of vegetables, with the exception of root vegetables, such as potatoes, sweet potatoes, and carrots, which tend to let go of their sugars more quickly.
- Whole fruits are fine, despite their sweet, seductive taste. Fruit juices also release their sugars fairly gradually, although they lack much of the fiber of the whole fruit.

Food Choices for Optimal Weight Control   123

- For extra slimming power, choose raw foods, such as cucumbers, apples, oranges, and pears, as often as possible. For reasons that have never been entirely clear, these foods can cause a powerful weight loss, even for people with stubborn weight problems.
- White bread, potatoes, and corn release their sugars rather quickly and can cause an overly strong insulin response that shuts down your calorie-burning machinery.

Similarly, foods that impair insulin's ability to work efficiently—fatty foods, primarily—end up forcing the body to make more and more insulin to compensate, which ends up promoting fat storage rather than fat-burning.

It bears repeating: this guideline applies only to those who have already dropped animal products and added oils from their diets and are still finding that their weight loss is too slow. They should choose legumes (beans, peas, and lentils), vegetables, fruits, whole grains, pasta, and plenty of raw foods and should avoid white bread, potatoes, and corn, and be careful to avoid animal products and other fatty foods.

One caution: Do not give this guideline more emphasis than it merits. If you are scrupulously avoiding white bread, but are still eating chicken, fish, or other fatty foods, you are wasting your time. That is not a formula for long-term success. Tuning up your insulin efficiency means, first, getting away from fatty foods, and second, choosing slow-release carbs.

MEAL PLANNING WITH THE NEW FOUR FOOD GROUPS

As you can see, whole grains, legumes, vegetables, and fruits are clearly the best choices for weight control. As we build them into a menu, tasty examples are everywhere. Spaghetti marinara, curried rice, or baked beans are all good choices. At a Mexican restaurant, skip the meat taco and choose a bean burrito. As noted above, good soup choices include minestrone, split pea, black bean, or bean chili made with chopped vegetables.

For breakfast, skip the usual eggs and bacon and have a big bowl of oatmeal with cinnamon and raisins, fresh fruit, and toast topped with jam instead of butter or margarine. You'll keep your leptin system working fine, and your LPL enzymes will have trouble finding any fat to store.

For lunch or dinner, a good starting plan is to cover about half your plate with a grain, such as brown rice, then add vegetables to another quarter or more of the plate. Put a legume, such as baked beans, black beans, peas, lentils, or chickpeas on the remaining quarter or less of your plate. If you are new to beans, start with small amounts, and be sure they are well cooked to avoid indigestion, and the same goes for cruciferous vegetables, such as broccoli, cabbage, or brussels sprouts.

You may wish to expand the vegetable portion, preferably including two vegetables in a single meal, such as carrots and spinach, and letting them replace some grains or beans.

Take a look in the recipe section. It has plenty of delicious and healthful choices. For breakfast, try French toast or sweet-potato waffles. Or have fat-free pancakes, hot oatmeal topped with fruit, or cold cereal with low-fat soy or rice milk. You may find that you feel more satisfied with a bit more protein in your breakfast. Most other cultures have found ways to build healthy vegetable protein sources into their breakfasts, in contrast to North Americans, whose breakfast protein comes from eggs, bacon, and sausage. Latin Americans count black beans among their breakfast foods, just as beans on toast are served in England, chickpeas in the Middle East, and grilled tofu in Japan. If these all sound alien, try the fat-free soy bacon and sausage substitutes that have become popular in recent years. They are excellent.

Speaking of meat substitutes, they also work well at lunch. A bologna or turkey sandwich is more healthful when made from the meatless "deli slices" sold at health food stores or in the produce department of larger groceries. They look and taste just like the meat varieties. Eat them with a bowl of minestrone, and you will not feel as if you are on a diet.

For dinner, try eggplant manicotti, vegetable curry, tamale pie, or portabellos from the recipe section. They are quick to prepare and delicious. If you make extra on Sunday, you can freeze it for later in the week. If you dine out, think international. Italian, Mexican, Chinese, Japanese, Thai, and Indian restaurants have endless choices, and all good restaurants now have vegetarian choices. If you go to a fast-food restaurant, favor those with salad bars or baked potatoes. At Mexican fast-food restaurants, try the bean burrito instead of a meat taco. Baked beans or spaghetti with tomato sauce are tasty and quick choices.

FOLLOW YOUR HUNGER AND SATIETY CUES

Just as our bodies know how much oxygen they need, they also know how much food they need. Your natural hunger and satiety cues are designed to start and stop your eating. However, these cues are thrown off when foods are packed with hidden calories or are so seductive that we want them when we are not at all hungry.

Meals based on the New Four Food Groups allow your natural hunger and satiety cues to work optimally. These nutritious foods are filling, but not overly so—they don't cause the after-meal pains that come when fatty meals slow down stomach emptying. They encourage normal digestion and elimination. If you have found yourself not the least bit hungry in the morning, but ravenous at night, these meals will help you get back into a healthier rhythm.

In general, it is not particularly important how many times a day you eat, or whether you have occasional between-meal snacks or not. As long as your foods fit the above guidelines, you'll be fine.

MAKING THE CHANGE

In our research studies, we use the food choices described in this chapter and that you'll see in the recipe section. Our participants sometimes go through an adjustment to the lighter taste, which takes a week or two, although most soon find it much easier than they would have guessed. The results are often astounding. Even though they have not changed their exercise routines and are not limiting the amount of food they eat, weight drops away, seemingly on its own and often dramatically.

As you can see, the most powerful foods are vegetarian, in fact, *vegan*— meaning free of all animal products. In our research, I have come to respect the power of these foods and would strongly encourage you to see what they will do for you. You will likely find that it is a whole lot easier to stick with a plant-based diet than a calorie-restricted diet. That is what University of Pittsburgh researchers found in a survey of young women who had tried both. The average person was able to stay on a calorie-restricted diet for only four months, while the median duration on a vegetarian diet was two years—and counting.[8]

As I emphasized in the preceding chapter, I would not suggest that you feel any need to immediately commit to this kind of "diet" forever (although that would certainly be a healthful choice). Rather, just try it out for three weeks. You will likely find that your tastes change, and the way you feel about food changes, too. If you like it, then you can stick with it.

But for three weeks, be consistent. Give your waistline a chance to trim down and, most important, give your tastes a chance to change, which will not happen if you keep reminding them of the seductions of fatty foods.

IF YOU HAVE DIABETES OR HIGH BLOOD PRESSURE

Whole grains, vegetables, fruits, and legumes have benefits other than weight control. They can lower your cholesterol level profoundly (many times more effectively than other diets), make diabetes improve or sometimes even go away, reduce blood pressure, cut cancer risk, reduce menstrual pains and premenstrual symptoms, and even help relieve arthritis and migraines. These effects are described in more detail in my previous books.

If you are taking medication for diabetes, high blood pressure, or any other condition, be sure to see your doctor as you begin the diet change, and let him or her adjust your dosages as necessary. Otherwise, using this diet along with your normal medication prescription will likely prove too strong a combination, and your blood sugar or blood pressure may drop too low. Let your doctor taper or discontinue your medication when the time is right.

For at least a few weeks, let yourself experience what it is like to be on as nearly perfect a diet as you can.

Similarly, to block premenstrual cravings, consistency is important. To make the craving-buster feature work, you have to follow the diet throughout the entire month, not just immediately before your period.

Now's the time to get started. In the next chapter, we will take a look at ensuring that we get complete nutrition along the way.

Chapter Nine

Complete Nutrition

N UTRITION IS confusing, often surprising, and some of the assumptions we make about what is best for our bodies turn out to be completely wrong. Many of us have no idea whether we are better off with more protein or less, with or without extra calcium, or with how much fat and which kind.

In this chapter, we'll look at how to ensure complete nutrition and see where typical diets go wrong. In many cases, research studies have turned conventional wisdom on its head. Let's look at the most common questions about complete nutrition.

HOW MANY CALORIES SHOULD I GET IN A DAY?

Your calorie needs should not be pegged to 1,200 calories, 1,800 calories, or any other arbitrary limit. The number of calories your body needs each day is determined by the amount of energy it uses up in powering your brain, your muscles, your basic metabolic functions, and whatever exercise you do. If your diet does not keep pace with your energy needs, something has to give. To keep in balance your body will sacrifice various activities to the extent it needs to.

Of course, when our bodies burn fat for energy, we are delighted. If our calorie intake is too low, however, other metabolic functions are sacrificed, too. Your metabolism will drop, and you will be more likely to binge, as we saw in chapter 3. If calorie intake drops very low, the menstrual cycle can really be disrupted, as if the body is deciding that this is not a good time to become pregnant.

Your best guides are your natural hunger and satiety cues, which tell you to eat when you are hungry and stop when you are full. Unfortunately, in affluent countries, food is so plentiful that adults often forget what the normal sensation of hunger actually feels like. They felt it as children playing or studying in school and eagerly waiting for lunchtime, but adults have a nearly constant supply of snacks in cabinets, snack machines, and desk drawers, and hunger is a thing of the past.

We often ignore our satiety cues, too. Even though our stomachs tell us we are full, our taste buds are still looking for more. The solution is twofold: First, if your diet has become completely erratic because of a very low calorie diet, let's bring things back to normal using the Rule of 10. As we saw in chapter 3, your diet should include—as a minimum—ten calories each day for every pound of your ideal body weight, to prevent metabolic slowdowns and binges.

Second, building your menu from whole grains, legumes, vegetables, and fruits allows you to eat larger portions than you could if you were including meats, dairy products, or other fiberless, calorie-dense foods. You'll soon find that your body knows when it's hungry and when it's full.

WEIGHT-LOSS DIETS USUALLY MAKE ME TIRED, SLUGGISH, AND CONSTIPATED. WHAT AM I DOING WRONG?

Typical weight-loss diets are so low in calories and vitamins that it is no wonder you feel out of sorts. Your energy intake is low, your metabolism slows down, and you will not feel like yourself. The key is to increase your food intake, using our Rule of 10, described above.

Very high protein diets, such as the Atkins diet, leave many people feeling drained because the healthful carbohydrates your body needs to build up your glycogen energy stores are omitted. The result is a massive loss of glycogen and a lack of energy. More on this in chapter 10.

Fatty diets can make you sluggish because fat actually increases the viscosity ("thickness") of your blood, making it more like oil and less like water. Fat also prolongs the stomach's emptying time after meals.

Constipation is a common problem in low-calorie diets and in high-protein diets. They have too little fiber to keep your digestive tract working properly. This is not just a nuisance. It is a warning sign. Over the long run, fiber-depleted diets increase your risk of diverticular disease, hemorrhoids, and even colon cancer.

HOW MUCH PROTEIN DO I NEED?

In decades past, nutrition authorities thought that the more protein in your diet, the better. In recent years, this idea has been set aside in favor of a much more moderate view of protein.

The first problem was that high-protein foods, such as chicken, beef, and eggs, also packed enough fat and cholesterol to raise the risk of heart problems and certain forms of cancer. However, other problems soon surfaced. The amino acid "beads" that make up animal protein molecules are high in sulfur, which harms the body's ability to hold on to calcium. They actually tend to leach calcium from the bones. Calcium passes from the bones into the bloodstream, filters through the kidneys, and is lost in the urine. Yale University researchers found that the more meat people eat in various countries, the more fractures they tend to have, a sign of the bone-weakening effect of animal protein.[1] Likewise, when researchers feed animal protein to volunteers and then test their urine later, they find that bone calcium ends up in their urine.[2] This occurs with the protein in red meats, poultry, fish, eggs, and dairy products. Two egg whites, for example, contain twelve grams of animal protein, a huge amount that no one needs. The bone-thinning disease osteoporosis is epidemic in Western countries, and animal protein is an important part of its cause.

These proteins are also hard on the kidneys, causing the kidney filter units to gradually wear out and die, which is why doctors treating kidney patients greatly restrict their protein intake. As you can imagine, weight-loss regimens that emphasize high-protein foods and use little carbohydrate are dangerous over the long run.

Plant proteins, on the other hand, appear to be free of these problems. The legume group (beans, peas, and lentils) is particularly protein-rich, and vegetables contain a fair amount of protein, too. Asparagus is 42 percent protein, as a percentage of calories. Broccoli is 40 percent protein.

These foods provide all the essential amino acid building blocks to make up the proteins in the human body. The old notion that you need to carefully combine or "complement" various plant foods to get adequate protein has been set aside. Both the U.S. government and the American Dietetic Association hold that, so long as your diet includes a normal variety of plant foods, you will easily get enough protein, even without any special combining.

HEALTHY HIGH-PROTEIN FOODS

(serving size: 1 cup, cooked)

	CALORIES	PROTEIN (GRAMS)	FAT (GRAMS)
LEGUMES			
Baked beans	235	12.2	1.1
Black beans	227	15.2	0.9
Chickpeas	285	11.9	2.7
Kidney beans	225	15.4	0.9
Lentils	231	17.9	0.7
Lima beans	217	14.7	0.7
Navy beans	259	15.8	1.0
Peas	134	8.6	0.4
Pinto beans	235	14.0	0.9
Split peas	231	16.4	0.8
SOYBEAN PRODUCTS			
Soybeans	298	28.6	15.4
Soy milk (Edensoy Extra)	140	10.0	4.0
Tempeh	330	31.4	12.8
Tempeh burger (1)	110	12.5	3.2
Tofu (firm)	366	39.8	22.0

Turn Off the Fat Genes

BREAKFAST CEREALS

All-Bran	213	12.0	1.5
Cheerios	89	3.4	1.4
Cornflakes	110	2.0	1.0
Cream of Wheat	153	4.4	0.5
Fruit & Fibre	182	5.6	2.0
Grape-Nuts	416	12.4	0.4
Oatmeal	144	6.0	2.4
Wheaties	99	2.7	0.5

OTHER GRAINS

Corn	178	5.4	2.2
Quinoa	159	5.5	2.5

VEGETABLES

Asparagus	44	4.6	0.6
Broccoli	46	4.6	0.4
Brussels sprouts	60	4.0	0.8
Swiss chard	36	3.4	0.2

FRUITS

Figs (10, dried)	477	5.7	2.2
Raisins (seedless)	450	4.8	0.8

SOURCE: Pennington JAT, *Bowes and Church's Food Values of Portions Commonly Used* (Philadelphia: Lippincott-Raven, 1998).

If you were looking for another reason to get your protein from whole grains, vegetables, and beans, rather than from meats, eggs, and dairy products, there is another risk you should know about. When animal products are heated in cooking, cancer-causing chemicals tend to form on their surface. Called *heterocyclic amines,* these carcinogens form as heat hits the creatine, amino acids, and natural sugars within the animal muscle. Grilled chicken is especially high in heterocyclic amines, with measured levels fifteen times higher than those in beef. But these chemicals form in all meats exposed to high cooking temperatures. Happily, they tend not to form in plant products. If you fry a hamburger or grill some chicken, carcinogens will form, but if you cook a veggie burger, it just gets warm.

It is possible to get too little protein, although this is quite uncommon. A person avoiding vegetables and legumes, and eating a very restricted diet of mainly grains and fruits, might run low on protein. On rare occasions, I have seen individuals whose hair is thinning and who have little energy, and these symptoms resolve when they bring beans and vegetables into their diet.

SHOULD I AVOID CARBOHYDRATES? PEOPLE SAY THEY MAKE YOU FAT

People in rural Japan, Thailand, China, and other Asian countries eat phenomenal amounts of carbohydrates in rice and starchy vegetables, and they are the slimmest people on earth. Not until Western eating habits bring meat, cheese, and similar high-fat, high-protein foods to these countries do weight problems become commonplace.

No, carbohydrates, such as rice, breads, and pasta, do not make you fat. But what you put on these foods—butter or margarine on bread, or greasy toppings on spaghetti—can easily add to your body fat. As we saw in chapter 4, there is no reason to avoid a potato if the real blame goes to the butter, sour cream, and melted cheese that smothered it with fat calories, nor should we blame the pasta when it is drowning in fatty Alfredo sauce.

Scientists have biopsied people's fat stores and found that virtually all of their fat has come from fat in the foods they have eaten, and almost none of it is produced from carbohydrate.

It is a mistake to avoid the vegetables, grains, beans, and fruits that provide healthy complex carbohydrates. If you do, your body will miss the nutrients it needs. If this causes weight loss, it is mainly because you are cutting out so many foods that your calorie intake has plummeted. In other words, you are starving and would see the same result from eliminating any major part of your diet. There is no reason to blame carbohydrates for weight problems. The true culprit is elsewhere.

Turn Off the Fat Genes

WHAT ABOUT IRON?

Iron is a double-edged sword. Your body needs a certain amount of iron so that your red blood cells can build hemoglobin to transport oxygen. You can get into trouble from having too little iron in your diet, but also from having too much. In excess, iron encourages the formation of free radicals, unstable molecules that form as a sort of waste product from the various chemical reactions that go on inside your body. Free radicals contribute to aging and can be the spark that starts the formation of cancer cells and artery blockages.

Iron comes from the foods you eat. Green leafy vegetables and beans are rich in a form of iron that is more absorbable when you need more iron and less absorbable when your body already has plenty. Foods rich in vitamin C—fruits and vegetables, that is—increase iron absorption. Dairy products slow down iron absorption, and some evidence suggests that coffee and tea might do the same thing. A person who tends toward anemia would do well to increase his or her intake of vegetables and legumes and avoid dairy products.

Meat products contain a form of iron called *heme iron*, which defies the body's ability to regulate its absorption. It passes through the digestive tract wall, into the bloodstream, whether you need more iron or not. Over the long run, this can boost the amount of iron in your body enough to increase the risks described above. You are better off getting your iron from plant sources.

If you are concerned about your iron status, your doctor should check it with special blood tests. A routine "blood count" (hemoglobin and hematocrit) is only a rough guide. If you are borderline low on either of these tests and feel perfectly fine, this is not a reason to diagnose anemia. In fact, from the standpoint of heart disease and general health, some evidence suggests that you are better off being slightly on the low side, rather than having too much iron. However, if your hemoglobin or hematocrit is distinctly below the norm, or if you are actually symptomatic—with low energy, for example—your doctor must investigate the cause of your low blood count and will do a few additional tests to check how much stored iron is in your body.

Menstrual flow does reduce iron levels slightly, but does not normally cause anemia. Doctors will usually look for other causes of blood loss. In adjusting your diet, resist the old, unhealthy advice to get iron from liver or meat. These foods can add a tremendous amount of fat and cholesterol to your diet. I was recently checking lab results on some research partici- pants. One man was consuming an average of 1,135 milligrams of choles- terol and over 100 grams of fat per day—an astounding amount of these unhealthy nutrients. His dietary records soon revealed the culprit—he had been eating lots and lots of liver, the fatty organ that actually produces most of the cholesterol in a cow's body. Yes, you'll get a lot of iron from it, but you'll also get a lot of what you don't want.

A sensible approach to keeping a healthy iron level is to have a diet rich in vegetables and legumes, and avoid dairy products. Follow your doctor's advice regarding any medical tests that may be recommended. Iron sup- plements are not usually necessary.

If your doctor finds that you have too much iron in your blood, by all means avoid the foods that drive even more iron into your body—i.e., meat products. The only way to quickly and reliably reduce the amount of iron in your blood is to donate blood, believe it or not. People who donate blood regularly tend to keep a much healthier iron level. Exercise also helps keep iron levels in bounds.

WHAT IS THE BEST WAY TO KEEP CALCIUM IN BALANCE?

For calcium, two issues are important: first, holding on to the calcium you have in your bones already, and second, including calcium in your diet to make up for natural losses.

Osteoporosis is a serious problem, particularly for Caucasian women after menopause. But it is not usually due to not getting enough calcium. Rather, it is a problem of abnormally rapid calcium *loss*. Calcium passes from the bones into the blood, filters through the kidneys, and is lost in the urine. The factors that affect this calcium drain have been identified, and they might surprise you:

- As we noted above, animal protein is a major culprit in the loss of calcium from bones, and avoiding animal protein com- pletely can cut calcium losses dramatically.[2]

- Sodium is aggressive at encouraging calcium loss via the kidneys. Cutting your salt intake to one to two grams per day can reduce your calcium needs substantially. To do this, look out for salt in canned goods and snack foods—often the amount is phenomenally high. Of foods prepared without added salt, dairy products are fairly high in sodium, although not so high as canned or snack foods.[3]
- Go easy on the caffeine. If you limit your caffeine intake to no more than two cups of coffee per day, your bones will have an easier time holding on to calcium.[4] If you drink more than this, use decaf, or try one of the coffee substitutes sold in health food stores.
- Don't smoke. Long-term smokers have 10 percent weaker bones, compared to nonsmokers. That 10 percent difference can spell a 44 percent increase in the risk of a hip fracture.[5]
- Regular physical activity helps keep your bones strong, while sedentary living weakens them.
- Your bones also like a little sun. As sunlight touches the skin, it turns on the natural production of vitamin D, which helps keep your bones strong. Brief periods of sun exposure regularly can give your body the vitamin D it needs.
- If you rarely see the sun, you will need a vitamin D supplement. Any common multiple vitamin containing five to ten micrograms (200–400 IU) provides a day's worth of vitamin D. Avoid higher doses unless prescribed by a physician.

You do need calcium in your diet. But don't depend on a high calcium intake to protect your bones if you are not controlling the calcium-depleting factors. Milk, in particular, is poor insurance against bone breaks, despite aggressive advertising of milk's supposed benefits.

The Harvard Nurses' Health Study tested the effect of milk on bones. The research team followed 77,761 women for twelve years and reported in 1997 that those who drank the most milk actually broke the most bones. Specifically, those who got the most calcium from dairy sources had nearly double the risk of hip fractures, compared to those who got little or no calcium from dairy products. When the statistics were adjusted for confounding factors—weight, menopausal status, smoking, and alcohol use—the relationship still held.[6] Yale researchers found much the same thing. Contrary to what one might expect, countries with higher calcium consumption actually have *more* hip fractures, not fewer.[1]

Some researchers speculate that milk drinking may actually increase

HEALTHY HIGH-CALCIUM FOODS (MILLIGRAMS)

Black turtle beans (1 cup, boiled)	103
Broccoli (1 cup, boiled)	94
Butternut squash (1 cup, boiled)	84
Chickpeas (1 cup, canned)	80
Collards (1 cup, boiled)	358
Corn bread (one 2-ounce piece)	133
English muffin	92
Figs, dried (10 medium)	269
Great northern beans (1 cup, boiled)	121
Kale (1 cup, boiled)	94
Mustard greens (1 cup, boiled)	150
Navy beans (1 cup, boiled)	128
Oatmeal, instant (2 packets)	326
Orange juice, calcium-fortified (1 cup)	270
Pinto beans (1 cup, boiled)	82
Soybeans (1 cup, boiled)	175
Spinach (1 cup, boiled)	244
Sweet potato (1 cup, boiled)	70
Swiss chard (1 cup, boiled)	102
Tofu ($\frac{1}{2}$ cup)	258
Vegetarian baked beans (1 cup)	128
White beans (1 cup, boiled)	161

SOURCE: Pennington JAT, *Bowes and Church's Food Values of Portions Commonly Used* (Philadelphia: Lippincott-Raven, 1998).

fracture risk by disrupting the body's natural calcium balance. I believe that the most we can say at this point is that milk does not protect against fractures, and that the key is to focus on factors that keep calcium in our bones: avoiding animal protein and excess sodium, staying physically active, getting some sun, and steering clear of tobacco and excess caffeine.

When you avoid the calcium depleters, you can keep strong bones with a fairly modest calcium intake. Of course, if you get little calcium, say less than 400 milligrams per day, you may not be giving your body the calcium it needs.

WHAT ARE THE BEST CALCIUM SOURCES?

The healthiest calcium sources are green leafy vegetables and legumes or, as some people say, "greens and beans." Green vegetables typically have calcium absorption rates of over 50 percent, compared with only about 32 percent for milk. Beans and bean products, such as tofu, are also rich in calcium. You don't need to eat huge servings of vegetables or beans to get enough calcium, but do include both in your regular menu planning.

If you are looking for extra calcium, fortified orange, apple, or grapefruit juices are good choices. They have 300 milligrams of calcium or more in an eight-ounce serving.

ARE THERE ANY SPECIAL CONSIDERATIONS FOR VEGETARIANS?

A low-fat, pure vegetarian (vegan) diet is the healthiest diet there is. In planning for complete nutrition, three simple issues often come up.

First, many vegetarians assume that, if they are not eating meat, they need to make up for it by consuming a fair amount of dairy products. In fact, they would do better to make up for the absence of meat with plenty of vegetables, beans, fruits, and whole grains.

There are several problems with dairy products:

- They have virtually no iron and can inhibit iron absorption, as noted above.
- The milk sugar lactose can encourage unpleasant digestive symptoms.

Complete Nutrition

- Milk proteins can trigger respiratory conditions, arthritis, migraines, and skin problems.
- The fat and calories in dairy products can lead to weight gain.
- Dairy products, whole or nonfat, have no fiber and no complex carbs. Nonfat milk derives its calories from lactose sugar (55 percent of calories) and animal proteins (about 40 percent of calories).

The dangers of a dairy-rich diet are evident in India. There, many people follow a traditional vegetarian diet, which overall is an excellent choice and could teach a great deal to Westerners. However, many Indians make up for the lack of meat with large amounts of milk, yogurt, cheese, and clarified butter. As a result, the population has a high prevalence of iron deficiency, compared to a country like China, which similarly consumes little meat but traditionally avoids dairy products. Some investigators have linked prostate and breast cancer to hormonal changes that result from milk drinking, and these, too, are more common in India than in China. While these health issues are being sorted out in research, there is every reason to steer clear of dairy products.

Second, don't assume that vegetable oils are slimming foods. Yes, vegetable oils are better for your heart than animal fats, but they are just as fattening as lard. "But I'm eating *good* fat," some people say, and true enough, we do need tiny traces of "good fats," which are found in beans, vegetables, and fruits. But if you're loading up your pasta with a layer of olive oil, or taking in fish oils for that matter, you might be surprised at how much "good fat" your body can load onto your thighs.

One nutrient does deserve a bit of planning, although it is a simple issue. Vitamin B_{12} is needed in tiny amounts for healthy blood and healthy nerves. It is not made by plants or animals; it is made by bacteria and other microorganisms. Long ago, when our ancestors plucked vegetables from the ground, there were likely to be traces of B_{12} in the soil, on plants, on their fingers, and even in their mouths, which provided traces of this vitamin. Modern hygiene has eliminated these sources, just as the modern workplace has all but eliminated the sun exposure that is the natural source of vitamin D.

It is easy to get adequate B_{12} from common fortified breakfast cereals, other fortified products, or any common multivitamin. The recommended dietary allowance is only two micrograms per day, with no danger from higher amounts. Spirulina, sold at health food stores, is not a reliable source.

Meat eaters get the vitamin B_{12} produced by fecal bacteria in the intestinal tracts of animals, which passes into the animals' tissues, but, unfortunately, they get a load of fat and cholesterol along with it.

Your body stores a good supply of this vitamin, so there is no reason to panic, but it is important to be sure to include a reliable source of vitamin B_{12} in your diet.

Chapter Ten

About High-Protein Diets

MANY PEOPLE in recent years have dabbled with high-protein diets in search of weight loss. The Atkins diet and its spin-offs (*Enter the Zone, Sugar-Busters,* and others) have encouraged readers to blame their weight problems on bread, potatoes, cookies, and cakes. By avoiding these carbohydrate demons, says Dr. Robert Atkins, people can dig into steak and eggs to their heart's content.

To frustrated dieters, it sounded halfway reasonable. Proteins are the nutrients that are supposed to make us strong; carbohydrates are what we feel guilty about. To dietitians, however, it was a nightmare. Diets loaded with meat, cheese, and eggs are known to harm the kidneys, weaken bones, increase cancer risk, and create all manner of health havoc.

Individual experiences have been mixed. In the first month or so on a high-protein diet, some people lose weight; others do not. Some find the diet palatable, at least for a few weeks, while others find their energy depleted, and they miss their normal breads, cereals, and other foods.

Unfortunately, those looking for research studies to settle whether high-proteins work have found virtually nothing to go on. In theory, there was little reason to think they would work well, but virtually no one has studied their effects in people with typical weight problems.

Here is the usual rationale for high-protein diets: The body normally fuels its activities with natural sugars (especially glucose), which come from complex carbohydrates. By removing nearly all carbohydrates—starchy vegetables, breads, pasta, rice, etc.—from the diet, the body has no choice but to burn fats that come either from foods or from body fat. In the process, the body produces ketones, chemicals that can be detected on the breath and in the urine.

In *Enter the Zone,* Barry Sears refocused the theme slightly by casting particular blame on insulin. As we saw in chapter 5, insulin excesses can contribute to weight problems. Since carbohydrates spark insulin release, they were viewed as evil.

Several things are wrong with these explanations. First, people in rural Asia consume enormous amounts of carbohydrate-rich rice and noodles, and they are the slimmest people on the planet. Not until they trade their traditional diets for meat and other Western tastes does weight gain begin. Moreover, in research studies, increasing carbohydrates at the expense of dietary fat does not cause weight gain at all. It does the opposite. In one study after another, it causes weight loss.

Some authors have blamed carbohydrates for weight problems because, as the notion goes, Americans are getting fatter, even though their fat intake has gone down in recent years, so carbs must be the culprit. This is a remarkably persistent myth. The truth is that the burgeoning obesity epidemic of the 1980s paralleled an increase, not a decrease, in fat intake. As noted in chapter 4, large U.S. food surveys showed that daily fat intake among adults increased from 81 grams in 1980 to 86 grams in 1991. The reason some suggest that fat intake must not have risen is that consumption of everything else went up, too, so the *percentage* of calories from fat was little changed. The truth is, fat intake hasn't gone down, it has gone up and remains by far the greatest contributor to weight problems.

This is not to say that cutting out all carbohydrates has no effect. Indeed, it can induce ketosis, in which the body is desperately trying to find fuel to power its activities. Eliminating carbohydrates also rapidly depletes the glycogen from your muscles, which is why such diets often cause fatigue. This also causes a temporary loss of the traces of water that surrounded each glycogen molecule, and that can be mistaken for a loss of fat. And since carbohydrates normally make up roughly half the calories people consume, eliminating foods rich in carbs knocks out a fair amount of calories, just as would eliminating any other major part of the diet.

The effect on insulin has turned out to be quite different from that predicted by Sears. In 1997, the *American Journal of Clinical Nutrition* published data on the insulin-stimulating effect of various foods, showing that high-protein foods often stimulate more insulin release than do carbohydrates. Most vexing to protein proponents was that fish produced a bigger insulin release than popcorn, and beef caused a bigger insulin release than pasta.[1]

The most powerful diet for taming insulin excesses is not a high-protein diet but a low-fat, fiber-rich, vegetarian diet. Rather than deny you carbohydrate-rich foods, this kind of healthful diet helps repair your body's ability to use them, so insulin excesses are avoided.

THE DANGERS OF HIGH-PROTEIN DIETS

High-protein diets present numerous health concerns. While they are loaded with fat and protein, they are low in vitamin C and B vitamins. The absence of fiber often contributes to constipation and other bowel problems.

A high intake of animal protein can be harmful to the kidneys, gradually damaging the nephrons, the kidneys' tiny filter units. The effect is not usually detectable over the short term, but becomes more serious the longer one adheres to the diet. Animal protein also encourages the loss of calcium from the bones, encouraging osteoporosis.[2]

Over the long run, a meaty diet increases the risk of several forms of cancer, particularly colon cancer. Part of the reason is that, as meats are cooked, amino acids, creatine, and natural sugars in the animal muscle tissue coalesce to form cancer-causing chemicals called *heterocyclic amines,* as we saw in chapter 9. These chemicals are part of the explanation for the threefold increase in colon cancer among frequent meat eaters, compared to those who generally avoid meats.[3]

Of concern to dieters, ketones pass from the bloodstream through the lungs and are easily detectable on the breath. To Atkins, this is a sign that the diet is working, but to dieters, it is a cause of self-consciousness.

About High-Protein Diets

THE FIRST RESEARCH STUDIES

On February 24, 2000, Robert Atkins, Barry Sears, and *Sugar-Busters'* Morrison Bethea faced off against Dean Ornish, John McDougall, and other advocates of low-fat and vegetarian diets in a morning-long debate sponsored by the U.S. Department of Agriculture. Most notable in the debate was the marked absence of scientific evidence favoring high-protein diets, in contrast to innumerable scientific studies establishing the slimming effects and other health benefits of low-fat and vegetarian diets.

Responding to the lack of data on high-protein diets, Duke University researcher Eric Westman recently conducted an uncontrolled study, paid for by Atkins himself, of a high-protein diet along with various nutrition supplements and regular aerobic exercise. The result was a weight loss of about one pound per week over twenty weeks, and a modest drop in cholesterol levels of 5 to 6 percent. Similarly, data from 319 patients at the Atkins Center in New York showed an average weight loss of about seventeen pounds in a year. Side effects in the Duke study included headache, fatigue, nausea, dizziness, constipation or loose stools, abdominal cramps, and bad breath.

A high-carb, low-fat, vegetarian diet is far better over the long run. Dr. Ornish's heart patients, treated with a low-fat, vegetarian diet and regular exercise, lost twenty-two pounds over a year, with a 23 percent drop in cholesterol levels and no troubling side effects at all.[4]

In summary, high-protein diets may cause some people to drop some pounds over the short term, but their weight-loss effect is no better than that of carbohydrate-rich diets, their effect on cholesterol is much more modest, and side effects are more common. It is possible they work for some people because cutting out *any* type of food leads to a drop in overall calorie intake.

High-protein diets present a Band-Aid approach to weight problems, and not a very effective one. Over the long run, low-fat diets rich in whole grains, legumes, vegetables, and fruits are much better at getting your body's insulin working again, trimming your waistline and cholesterol level, and maximizing health.

Chapter Eleven

Children and the Fat Genes

WHEN NEW babies greet the world, they begin a time of rapid growth. As the months go by, facial features become more defined, fingers become more dexterous, and the brain and nerves weld connections that allow the child to crawl, stand, and walk. Although the number of fat cells remains essentially constant, the cells themselves shrink as baby fat is trimmed away.

By the preschool years, however, the fat genes are already at work for some kids. Chromosome 8 will have built plenty of lipoprotein lipase (LPL), the enzyme we met in chapter 4. As the child walks through the school lunch line, LPL lies in wait for the inevitable fat particles that will soon arrive in the bloodstream. It extracts the fat and passes it into fat cells. Insulin, pressed into overtime by the influx of too much protein and sugar, ensures that it is stored, rather than burned. The fat cells grow bigger and more numerous.

If the growing child later reacts to real or imagined overweight with a crash diet—as a frightening number of teens do—the low calorie intake quickly disables leptin, the hormone that is supposed to keep appetite in check and maintain a quick metabolism.

Not every chubby baby will become an overweight teenager or adult. But those who do will face real challenges. They will walk the tightrope between ever-worsening weight problems and harmful diets (even eating disorders), confronting a higher than normal risk of heart problems, joint problems, gallbladder problems, and even some forms of cancer.

In this chapter, we'll take a look at how to quiet the genes that encourage a child's tendency to store fat.

PREVENTING WEIGHT PROBLEMS BEFORE BIRTH

Paradoxically, a mother can help prevent weight problems in her child by being sure she eats *enough* during her pregnancy, as a huge and unfortunate wartime "experiment" proved. Beginning in December 1944, massive food shortages in Holland limited food rations to just 400 to 800 calories per day until the Allied liberation the following April. Babies in the first trimester of development when the famine hit had normal weights at birth. But in adulthood the females in this group began putting on extra weight. By age fifty, they were seventeen pounds heavier, on average, than other women of similar age. Males seemed not to show this effect, for reasons that are not clear. Babies in the middle or final trimesters of pregnancy when the famine arrived were smaller at birth and tended to stay slimmer throughout life.[1]

The lesson of this experience is this: a baby's body sets its energy balance early in pregnancy. Skimping on food can make for a miserly metabolism that tends to store calories. If, on the other hand, an expecting woman eats normal amounts, the child has no special biological tendency toward weight gain.

A pregnant woman needs about 300 calories per day more than she otherwise would, and they should come from vegetables, fruits, grains, and legumes (beans, peas, and lentils), which provide plenty of vitamins and minerals as well. Her obstetrician will add prenatal vitamins and should check whether she needs an iron supplement based on blood tests at the beginning and middle of pregnancy.

HOW OVERWEIGHT BEGINS

A newborn baby is strongly attracted to sweet tastes, which are found in mother's milk and in the fruits that will be among the infant's first solid foods. While a "sweet tooth" may have a bad name, these natural sources of nourishment do not cause childhood weight problems.

PREDICTING OBESITY

The real culprits in obesity are the traces of fat picked up by the same lipoprotein lipase (LPL) that stores fat in adults. LPL lies in the blood vessels that course through the fat tissue, and it greedily extracts fats or oils from the bloodstream and adds them to body fat. A typical kid's diet gives LPL plenty to work with. A glass of whole milk is 49 percent fat. Chicken McNuggets are 59 percent fat, with more than twenty grams of it dripping out of a single serving. A hamburger packs in ten grams, a modest serving of potato chips another ten, and a hot dog fifteen.

Every last gram is absorbed by the child's body, and much of it will add to body fat. Even skinless chicken breast, which well-meaning parents may imagine to be reasonably lean, delivers several times the fat found in rice, vegetables, fruits, or beans. The children who put on body fat early in life are those whose diets are rich in meats, fatty dairy products, and fried foods, which keep their LPL occupied.[3]

Many pediatricians and parents labor under the myth that children need fairly large amounts of fat in their diet for brain development. The truth is that, after infancy, they need only rather small amounts: 10 to 15 percent of calories is sufficient. More important, the fats they need are those found in vegetables, beans, and certain plant oils, not the saturated fats prevalent in meats, cheese, and milk.

While some parents blame carbohydrates for their own or their children's weight problems, too many slices of bread or too many servings of potatoes, beans, rice, or spaghetti are not what's making kids gain weight.

Researchers have found that kids who gain weight tend to be those who *avoid* carbohydrate-rich foods in favor of fats.[4]

Once a child is overweight, any attempt to lose weight will be stymied if there is a constant supply of calories from any source. Sugars, for example, contain less than half the calories of fats, ounce for ounce, and are not the cause of most weight problems. However, a child "treated" to an endless array of snacks at home will have a tough time burning off fat. An influx of sugar or protein sparks the release of insulin, which inhibits fat cells from burning fat and encourages fat storage. Many sweets also contain plenty of fat, which makes fat stores grow even more.

THE CONSPIRACY OF GENES AND THE ENVIRONMENT

As a result of genetic programming favoring the sweet taste of mother's milk and fruit, a child's tastes can easily be seduced by *anything* sweet. Chocolate, sugar candies, ice cream, and sodas are instant hits, so much so that we even think of them as natural for kids, forgetting that, from the standpoint of human history, they are all relatively recent inventions.

In the preschool years, children's eating habits are similar to those of their parents, especially their mothers.[5] Soon, however, their preferences mimic those of their friends and schoolmates. Regrettably, school lunch programs often serve lunches you would never provide at home, often high in fatty meats and dairy products.*

Fast-food companies aggressively target children. While the convenience and low prices of these products appeal to many parents, it is critical to be selective, favoring salad bars over burgers, baked potatoes over fried chicken, bean burritos over meat tacos, etc.

GOOD NUTRITION FOR KIDS

Good nutrition starts with breast-feeding, if at all possible. Mother's milk contains exactly the right nutrients for optimal growth in the first year or

---

*Healthful recipes and educational materials for school lunch programs and other institutional settings, such as hospitals or college cafeterias, are available from the Physicians Committee for Responsible Medicine, 5100 Wisconsin Avenue, Suite 404, Washington, DC 20016, 202-686-2210, www.pcrm.org.

two of life. It is rich in the fats that are essential for brain development and rich in natural sugars that support energy and growth, and it confers advantages that last a lifetime. Breast-fed babies have better immune defenses, have less risk of diabetes, and gain three to five IQ points, compared to formula-fed babies.[6] Breast-fed babies also have more rapid maturation of visual and motor systems and fewer behavioral problems. Cow's milk formulas have a somewhat different balance of fats and a very different sugar-protein balance, favoring calves, not human babies.

The foods that help children avoid weight problems are the same ones that help adults trim their waistlines. They are low in fat and do not cater to LPL's penchant for fat storage. They have a low glycemic index and thus do not stimulate any overrelease of insulin that can promote fat storage. They include the following four groups:

- Vegetables: carrots, sweet potatoes, green beans, etc.
- Legumes: baked beans, garden peas, lentils, etc.
- Fruits: apples, bananas, oranges, pears, etc.
- Whole grains: brown rice, whole grain bread or noodles, corn, etc.

Also, be sure to add a source of vitamin B_{12}, such as a daily multivitamin or fortified cereals or soy milk.

Do not skimp on portion sizes. In general, it helps to be as generous with vegetables, fruits, and legumes as the child's tastes will permit, as these foods are richer in many vitamins than are typical grains. It is also important to bring variety into the diet. While this is rarely a problem in Western countries, diets in developing countries are sometimes restricted to a few items.

Children who grow up with rice dishes, baked beans, lentil soup, spaghetti marinara, whole-grain breads, fresh fruit, selected vegetables, and other healthy foods automatically prefer these tastes. These foods provide complete nutrition for growing bodies, without the excess fats that are on far too many children's plates. Because these foods help avert weight problems, they also prevent the dangerous reactions to overweight many teens stumble into: severe calorie restrictions, anorexia, and bulimia.

In planning meals, one truth is eternal: Most children will never be big on broccoli or spinach. Not only is a kid's sweet tooth in overdrive, due to genetic programming, but three out of four children are genetically programmed to be sensitive to bitter tastes, as we saw in chapter 2. For them, bitter vegetables are totally unpalatable.

They will gradually develop a taste for vegetables as the aversion to bitter tastes melts away in adulthood. Meanwhile, there is no need to push kids to eat foods they do not care for. Simply provide the vegetables they *will* eat—carrots, green beans, sweet potatoes, peas, or other nonbitter choices. You may also find that vegetables added to soups or stews in unrecognizably small bits will not trip the child's bitter-taste sensors.

Animal products (meats, dairy, and eggs) should be avoided completely, and cooking oils should be kept minimal. In general, they provide way too much fat and too little nutrition. They have little or no beta-carotene, folic acid, vitamin C, fiber, or complex carbohydrate, and yet they are so high in calories that children filling up on these products will miss out on these vital nutrients.

Fatty foods can easily insinuate their way into a child's palate. Meats, for example, are not particularly attractive to small children at first tasting. However, the child's brain soon associates meat's taste with its considerable load of calories and fat and comes to favor it as an energy-dense food. The same thing happens with cheese, greasy french fries, and other calorie-rich products. The brain assesses each food, and those that are calorie-dense become favorites.

This was a useful ability when early humans had to forage for food and calorie-dense foods were rare. Nowadays, children in Western countries are drowning in calorie-dense foods, and their natural tendency to detect and favor them has become a major liability. The best thing parents can do for their children is to avoid starting them down this road by keeping such foods out of the home.

Happily, food manufacturers now provide healthful versions of burgers, hot dogs, and luncheon slices, all made from soy and wheat derivatives, tasting very much like the real thing. Take a look in the produce section or at any health food store.

Dairy products are not necessary or even advisable. Well-meaning parents often push cow's milk on children, buying into the myths that have come from relentless advertising. A child drinking three glasses of milk in a day swallows twenty-four grams of fat (more than 60 percent of which is artery-clogging saturated fat) and 450 calories.

The theory behind drinking milk in childhood is, of course, that it builds bone density, so that when the bone-thinning osteoporosis begins later in adulthood, the extra reserve of calcium can be drawn from. This has not been supported by good research. Over the long run, milk drinkers get no protection at all. Indeed, the Harvard Nurses' Health Study found that those who got the greatest amount of calcium from dairy sources had approximately double the risk of hip fractures, over a twelve-year follow-up period, compared to those who got little calcium from dairy products.[7]

Also, it pays to avoid stocking the pantry with sodas, cookies, and other unhealthful snacks. Nonetheless, there is no reason to prohibit healthy between-meal snacks, such as fresh fruit or baked goods with a reasonably low fat content.

MODEL HEALTHY EATING AT HOME

When a child is overweight, it is a warning sign for the entire family. The child may simply be the first to show the effects of a diet that is likely to lead to heart disease, diabetes, high blood pressure, and of course, overweight for other family members down the road. Rather than stigmatizing the child with an individualized diet or exercise plan, the whole family should join in with healthy habits. They will benefit, too.

Parents who build their diets from chicken, beef, milk, and french fries, or who keep their shelves stocked with endless supplies of soda and candies, are playing with fire. They are likely to gain weight themselves, and their children will follow suit.

When I was a child, my father was trying desperately to quit smoking, and eventually he succeeded. In the meantime, he was always careful not to smoke at home or in front of his children. He did his best to avoid encouraging us to follow him into this dangerous habit. Parents who stray from healthful diets themselves should take great care not to lead their children along the same path. Just as a smoking parent bathes his or her children in tobacco smoke, parents who eat unhealthful foods may end up fattening their kids, too, and teaching them dangerous tastes.

The risks are not only measured on the scale. Children who learn to eat a meaty, cheesy diet often develop the first stages of artery blockages before they finish grade school. The first place that artery blockages form is in the arteries that branch off the aorta to nourish the lower back. They begin in the preschool years, and by age twenty, about 10 percent of people in Western countries have at least one advanced blockage in one or more lumbar arteries. The result is a loss of oxygen and nutrients to the spine.[8] As the vertebrae and disks lose their blood supply, they are unable to bounce back from the wear and tear of day-to-day life, and increased risk of chronic lower-back problems is the inevitable result.* The same kind of artery blockages gradually form in the heart, leading to heart attacks later in life, and in the arteries to the brain, leading to strokes. While we associate these dangers with later life, the damaging artery changes often begin in childhood.

Foods eaten in childhood influence health in many other ways. Children brought up on vegetarian foods have significantly less risk of cancer, diabetes, hypertension, gallstones, and other conditions later in life.[9]

EXERCISE IS IMPORTANT, BUT . . .

Exercise is as important for children as for anyone else. Movement helps keep you slim. Researchers in Cambridge, England, studied infants as they grew and found that they could predict which would have the most body fat at their first birthday simply by measuring their spontaneous physical activity—moving, fidgeting, walking, or crawling—at three months of age.[10]

Unfortunately, modern life conspires against exercise. Computers and televisions fix children in position. However, there is no reason to push children into heroic amounts of exercise or to make them feel guilty for being sedentary. Just strap on your own sneakers and show them how it's

---

*For a detailed discussion of the role of diet in back pain, see Barnard ND, *Foods That Fight Pain* (New York: Harmony Books, 1998).

Turn Off the Fat Genes

done. The most powerful predictor of activity levels in children is the activity level of their parents. Partly, this is the effect of genetic factors, as we saw in chapter 6. But it is also the result of parents involving children in their own activities.

A study of one hundred children and their parents in Framingham, Massachusetts, showed that if Mom is generally physically active—taking part in sports, exercise, or just staying on the go—her child is twice as likely as other kids to be fairly active, too. When Dad is more physically active than average, the child is nearly four times more likely to have the same trait, compared to other kids. And when both parents were active types, the child was six times more likely to follow suit.[11]

TV DINNERS, AND LUNCHES, AND BREAKFASTS . . .

Television makes kids fatter. Not only does it rob them of physical activity, it also encourages extra snacking. It's easy to chomp on a turkey sandwich and drink a glass of milk while you're watching television, but almost impossible to do so when you're riding your bike or playing football.

In a community experiment, third- and fourth-grade students in a San Jose, California, grade school were asked to watch no television and play no video games at all for ten days—challenging as that may have been for many of them—and after that to limit their TV and video-game use to seven hours per week. Seven months later, they had gained less body fat than children at a second school who were left to their own devices with TV and video games.[12]

Despite its value, exercise cannot undo the effects of a bad diet.[3] After exercising for an hour, a child can sit down at a lunch table and easily consume far more calories than were burned in the exercise. The best advice is for the family to follow the healthiest possible diet, and to add a regular exercise program to it.

HELPING CHILDREN STAY HEALTHY

The eating habits that are now common in North America and Europe—emphasizing meats, dairy products, and fried and snack foods—do not bring out the best in children. On the contrary, they encourage any genetic tendency toward weight gain to express itself abundantly. At a time when we know more about nutrition than ever before and people in Western countries are able to select any foods they wish, our children are more out of shape than ever before. We can change that, and easily, too. The answer lies in changing the menu. A diet based on healthful foods—plenty of fresh vegetables, fruits, whole grains, and legumes, chosen according to the child's tastes—allows the emergence of the best of health.

Chapter Twelve

Getting
Started

ENOUGH THEORY. It's time to put what we've learned into practice. Let's move beyond our scientific studies and pick up our knife and fork.

If your tastes have been programmed by genes and experience to prefer fatty or sugary foods, we can readjust them. Even strong genetically determined tastes can be dramatically changed, if you choose to do so. You can also bring your appetite back into bounds if the amount of leptin in your bloodstream or other aspects of your appetite-regulation system have been skewed by diets, binges, or chronic overeating.

If your LPL enzymes are still busily tucking away fat—as they are for most people—we'll select foods that politely frustrate them. And if your body does not deal well with carbohydrates, we will employ the same diet changes researchers use to repair this problem.

In the following section are recipes designed to boost the genes for weight loss and counter those that encourage weight gain. This is a remarkable set of tools. These foods will not only delight your biology—they are rich in vitamins, minerals, and fiber—but they are also designed to be tasty and quick to prepare.

In our research studies, as our participants begin healthy diets, we maintain a spirit of exploration and fun. Some recipes will turn out to be absolute delights, and a few may be duds. That's okay. It's all part of experimenting. Let me encourage you to enjoy this same sense of testing, trying, and experimenting as you go about solving your weight problem. Leaf through the menus and recipes, and try some out. You'll find that family members who are already at their desired weight will love them as much as those aiming to lose some pounds.

The best way to proceed, as I mentioned in chapter 7, is to plan a menu change for a short period—I prefer three weeks—and during that time to follow as perfect a diet as possible. This will give you a clean break from unhealthy tastes, let you learn new ones, and show you what it is like to have as perfect a menu as you can have.

Our adventure leads us not only into new tastes, it also brings us a whole new body. When we begin a new research project, our participants never know quite where they will end up. Those who expected continuing difficulties losing weight have often been stunned by what a breeze it can turn out to be. With a new understanding of how your genes work, you can now choose foods intelligently and enjoy the changes that are in store for you.

Let me also encourage you to be attuned to new information as the nutrition world continues to evolve. If you would like to learn about new research findings as they become available or about efforts to promote better health and nutrition, you may wish to join the Physicians Committee for Responsible Medicine, a nonprofit organization I formed in 1985. Whether you are a doctor or a layperson, our magazine, *Good Medicine*, and our Web site—pcrm.org—will bring you the latest news and ways you can help. You can reach us at 5100 Wisconsin Avenue, Suite 404, Washington, DC 20016, by phone at 202-686-2210, or at pcrm@pcrm.org.

I wish you the very best of health and success in your new endeavor.

Part Three

Menus

and

Recipes

The
Menu
Plan

GETTING STARTED IN CHANGING YOUR EATING HABITS

A S Y O U ' V E already discovered, permanent weight loss involves selecting foods that signal your body to burn calories rather than store them. This section provides you with cooking tips and information as well as menus and recipes that will enable you to make these foods a part of your everyday life.

Learning to eat differently is a matter of establishing new habits. With a little practice, these habits, which may seem unfamiliar at first, will become second nature. This section guides you, step by step, as you begin your new eating practices.

The section includes tips to help you get started with menu planning and shopping, followed by a discussion of ingredients that may be new to you. "Almost Instant Meals and Snacks" is a list of handy instant meal and snack ideas; "Menus for Two Weeks" gives you ideas for assembling the recipes in this book for complete, satisfying meals.

The recipes in this section are, for the most part, quick and easy to prepare. They utilize ingredients that are familiar and easy to locate in most grocery stores. Many of the recipes are for healthful versions of familiar foods, and each recipe includes nutritional information.

HOW TO USE THIS CHAPTER, STEP BY STEP

1. SELECT A MENU

You can save time and money by planning and shopping for a week at a time. One trip to the store means a lot less time looking for parking and standing in line. In addition, by planning ahead you'll spend less on impulse items and instant meals. The easiest way to get started is to follow the menus presented in this chapter. They will help you utilize the recipes in this book, as well as other foods, to prepare healthful and satisfying meals. Once you have developed confidence with these menus, you can begin customizing them to suit your needs and lifestyle.

2. PURCHASE THE INGREDIENTS

Use your menu to determine the ingredients that you will need. Make a shopping list of those ingredients you don't already have on hand. Add staples such as breakfast cereals, breads, prepared and instant foods, beverages, fruits, and vegetables. This list will enable you to purchase all the food you need for a week with a single shopping trip.

3. PREPARE THE FOOD

With your menu and ingredients on hand, you will find that you are able to prepare satisfying meals with minimum effort. You will notice that most of the recipes in this book provide six to eight servings. As a result, you will probably have food left over that can be used to provide one or more extra meals. In this way, the menu you create may actually provide meals for more than a week with no additional shopping, planning, or cooking!

Another time-saver is to serve the food you've prepared in a slightly different form each time. In this way, you can have maximum variety with minimum of preparation. The Quick Bean Dip (page 198) is a good example. As the recipe indicates, it may be served as a dip with vegetables or as a burrito filling. By adding a bit more water, it may be used as a sauce for potatoes or vegetables, and by adding even more water, or vegetable stock and a few chopped green onions, it makes a delicious soup.

Foods that take a bit of time to cook can be prepared in large enough quantities to provide for several meals. Brown rice is a good example.

Turn Off the Fat Genes

Once cooked, it can be easily reheated in a microwave or on the stove top and served as a side dish with a variety of recipes. It can also be added to soups and stews, or used as a filling in a burrito or wrap.

TIPS FOR CUTTING THE FAT

Foods that are high in fat are also high in calories. In addition to causing unwanted weight gain, a high-fat diet increases your risk for heart disease and several forms of cancer. By switching to a plant-based diet you will reduce your intake of fat considerably. The following tips will help you reduce your fat intake even further.

- Choose cooking techniques that do not employ added fat. Baking, grilling, and oven-roasting are great alternatives to frying.
- "Sauté" in a liquid such as water or vegetable stock whenever possible. Heat about ½ cup of water in a skillet (preferably nonstick) and add the vegetables to be sautéed. Cook over high heat, stirring frequently, until the vegetables are tender. This will take about five minutes. Add a bit more water if necessary to prevent sticking.
- Add onions and garlic to soups and stews at the beginning of the cooking time so their flavors will mellow without sautéing.
- When oil is absolutely necessary to prevent sticking, lightly apply a vegetable oil spray. Another alternative is to start with a very small amount of oil (1 to 2 teaspoons), then add water or vegetable stock as needed to keep the food from sticking.
- Use nonstick pots and pans to allow foods to be prepared with little or no fat.
- Choose fat-free dressings for salads. In addition to commercially prepared dressings, seasoned rice vinegar makes a tasty fat-free dressing straight out of the bottle.
- Avoid deep-fried foods and fat-laden pastries. Check your market for low-fat and no-fat alternatives.
- Replace the oil in salad dressing recipes with seasoned rice vinegar, vegetable stock, bean cooking liquid, or water. For a thicker dressing, whisk in a small amount of potato flour.

- Sesame Salt (page 207) and Sesame Seasoning (page 208) make delicious low-fat toppings for grains, potatoes, and steamed vegetables. Fat-free salad dressing may also be used as a topping for cooked vegetables.
- Applesauce, mashed banana, prune puree, or canned pumpkin may be substituted for fat in many baked goods with a bit of experimentation.

ALMOST INSTANT MEALS AND SNACKS

Keep a bag of prewashed salad mix and a jar of fat-free dressing on hand.

Add some canned kidney beans or garbanzo beans for an almost-instant salad.

Baby carrots are available in most markets and make a delicious and convenient snack.

Fresh soybeans (edamame) make a delicious snack or meal addition (see recipe on page 269).

Ramen soups are quick and delicious. Add some chopped fresh vegetables for a great vegetable soup.

Keep a selection of vegetarian soup cups on hand. These are great for quick meals, especially when you're traveling.

Burritos are quick to make and very portable. They can be eaten hot or cold. Simply spread some fat-free refried beans or fat-free refried black beans on a fat-free flour tortilla. Add some prewashed salad mix and a bit of salsa for a delicious and satisfying meal.

Mix fat-free refried beans with an equal amount of salsa for a delicious bean dip. Serve with baked tortilla chips.

Rice cakes and popcorn cakes come in a variety of flavors and make great snack foods. Spread them with fruit preserves or bean dip.

Frozen grapes make a wonderful summertime snack. Simply remove them from the stems and freeze them, loosely packed, in an airtight container.

Frozen bananas make a delicious, creamy frozen dessert. Simply peel the banana (insert a popsicle stick into the end if you like) and freeze it on a tray. Once frozen, wrap it in plastic wrap.

Pita Pizzas (page 235) are quick and easy to make. Serve them with Three Bean Salad (page 211).

Drain garbanzo beans and spoon into a piece of pita bread. Top with prewashed salad mix and fat-free salad dressing for a quick sandwich.

Heat a fat-free vegetarian burger patty in the toaster oven. Serve it on a whole grain bun with mustard, ketchup, barbecue sauce, and lettuce. Add sliced red onion and tomato if desired.

Keep baked potatoes in the refrigerator. For a quick meal, heat a potato in the microwave and top it with fat-free vegetarian chili and salsa.

Check your natural food store for fat-free vegetarian cold cuts. These make quick and easy sandwiches.

MENUS FOR TWO WEEKS

DAY 1

BREAKFAST

cold cereal or hot Multigrain Cereal
(page 179)
soy milk or rice milk
fresh fruit
herb tea

LUNCH

Pita Pizzas (page 235)
California Waldorf Salad (page 221)
Beets in Dill Sauce (page 248)

DINNER

Mexican Skillet Pie (page 283)
Gazpacho (page 244)
Corn Bread (page 192)
Fresh Lemon Curd (page 299)

DAY 2

BREAKFAST

French Toast (page 191)
Corn Butter (page 200)
maple syrup or spreadable fruit
fresh fruit
herb tea

LUNCH

Quickie Quesadillas (page 234)
Autumn Stew (page 242)
green salad with fat-free dressing

DINNER

No-Meat Loaf (page 286)
Mashed Potatoes and Gravy (page 252)
Fresh Broccoli Salad (page 210)
Gingerbread (page 296) with Pineapple Apricot
    Sauce (page 204)

DAY 3

BREAKFAST

Whole Wheat Pancakes (page 185)
Corn Butter (page 200)
maple syrup or spreadable fruit
fresh fruit
herb tea

LUNCH

Portuguese Kale Soup (page 238)
whole-grain bread or roll
Rainbow Salad (page 215)

DINNER

Stuffed Winter Squash (page 287)
Ratatouille (page 254)
Always Great Brown Rice (page 270)
Butterscotch Pudding (page 297)

DAY 4

BREAKFAST

Oatmeal Waffles (page 189)
fresh fruit or spreadable fruit
herb tea

LUNCH

Summer Vegetable Stew (page 247)
Garlic Bread (page 193)
green salad

DINNER

Potato Enchiladas (page 281)
Quick Black Bean Chili (page 261)
South of the Border Salad (page 219)
Strawberry Freeze (page 292)

DAY 5

BREAKFAST

Buckwheat Bananacakes (page 187)
Corn Butter (page 200)
fresh fruit
herb tea

LUNCH

VegiBurgers (page 227)
Potato Salad (page 215)
watermelon

The Menu Plan

DINNER

Polenta Pizza (page 276)
Broccoli with Sesame Salt (page 256)
Antipasto Salad (page 213)
Poached Pears with Butterscotch Sauce
   (page 302)

DAY 6

BREAKFAST

Breakfast Scramble (page 183)
whole-grain toast with Corn Butter (page 200)
Strawberry Sauce (page 203) or spreadable fruit
herb tea

LUNCH

Vegetarian Reuben Sandwich (page 228)
Texas Caviar (page 210)
green salad with fat-free dressing

DINNER

Lasagne Roll-ups (page 291)
Garlic Bread (page 193)
green salad with fat-free dressing
Banana Bundt Cake with Date Butter Frosting
   (pages 305 and 306)

DAY 7

BREAKFAST

Muesli (page 182) or Quick Breakfast Pudding
   (page 182)
soy milk or rice milk
fresh fruit
herb tea

Turn Off the Fat Genes

LUNCH

Burritos Supremos (page 234)
Mexican Corn Chowder (page 243)
baked tortilla chips with Zuccamole (page 198)

DINNER

Portabellos and Red Pepper Wraps (page 233)
Spicy Thai Soup (page 239)
Ginger Noodles (page 271)
Carrots in Orange Sauce (page 255)
Ginger Peachy Bread Pudding (page 301)

DAY 8

BREAKFAST

cold cereal or Rolled Grain Cereal (page 180)
soy milk or rice milk
fresh fruit
herb tea

LUNCH

Thai Wraps (page 230)
Yam and Corn Chowder (page 240)
green salad with fat-free dressing

DINNER

Eggplant Manicotti (page 277) with Simple
  Marinara (page 202)
Zucchini Corn Fritters (page 274)
Cucumbers with Creamy Dill Dressing
  (page 209)
Pumpkin Spice Cookies (page 304)

The Menu Plan

DAY 9

BREAKFAST

Multigrain Pancakes (page 186)
spreadable fruit or maple syrup
fresh fruit
herb tea

LUNCH

Garbanzo Wraps (page 232)
Sweet and Sour Cucumbers (page 218)

DINNER

Yves Hot Dogs
Baked Beans (page 265)
corn on the cob
Chocolate Pudding (page 298) or Carob Mint
    Pudding (page 299)

DAY 10

BREAKFAST

Cornmeal Flapjacks (page 188)
spreadable fruit or maple syrup
fresh fruit
herb tea

LUNCH

California Nori Rolls (page 236)
Spicy Thai Soup (page 239)
Ginger Noodles (page 271)

DINNER

Simple Pasta Supper (page 278)
Summer Succotash (page 249)

Turn Off the Fat Genes

green salad with fat-free dressing
Fresh Peach Crisp (page 294)

DAY 11

BREAKFAST

Breakfast Teff (page 181)
soy milk or rice milk
fresh fruit
herb tea

LUNCH

Quick & Creamy Beet Soup (page 242)
Spinach Salad with Curry Dressing (page 212)
whole-grain bread

DINNER

Missing Egg Sandwich (page 226)
Hoppin' John Salad (page 220)
Apple Crisp (page 295)

DAY 12

BREAKFAST

whole-grain toast or bagel
Corn Butter (page 200) or spreadable fruit
fresh fruit
herb tea

LUNCH

Chili Beans (page 264)
Bulgur (page 272)
green salad with fat-free dressing

Ensalada de Frijoles (page 214)
Cranberry Corn Bread (page 306)

DAY 13

BREAKFAST

Sweet Potato Waffles (page 190)
spreadable fruit or maple syrup
fresh fruit
herb tea

LUNCH

Red Pepper Hummus (page 197)
pita bread
baby carrots
Tabouli (page 216)

DINNER

Skillet Scalloped Potatoes (page 251)
Split Pea Soup (page 246)
Gingerbread Cookies (page 303)
Tapioca Pudding (page 300)

DAY 14

BREAKFAST

hot or cold cereal
soy milk or rice milk
fresh fruit

LUNCH

Garbanzo Salad Sandwich (page 225)
Couscous Confetti Salad (page 217)
Fruit Gel (page 296)

Turn Off the Fat Genes

French Green Lentils (page 263) or Red Lentil
   Curry (page 262)
Vegetable Curry (page 285)
Couscous (page 273)
Cucumbers with Creamy Dill Dressing (page 209)
Carob Mint Pudding (page 299)

FOODS THAT MAY BE NEW TO YOU

The majority of ingredients in the recipes are common and widely available in grocery stores. A few that may be unfamiliar are described below.

agar—a sea vegetable used as a thickener and gelling agent instead of gelatin, which is an animal by-product. Available in natural food stores and Asian markets. May also be called agar agar. Agar comes in several forms: powder, flakes, and a large stick. The powder is easiest to use. If only flakes are available, you will have to adjust the amount used, as flakes are not as concentrated. Use approximately 1½ tablespoons of flakes to substitute for each teaspoon of powder.

arrowroot—a natural thickener that may be substituted for cornstarch.

balsamic vinegar—a mellow-flavored wine vinegar that is delicious in salad dressings and marinades. Available in most food stores.

barley flour—can be used in baked goods in place of part or all of the wheat flour for a light, somewhat crumbly, product. Available in natural food stores and many supermarkets.

Boca Burgers—a fat-free vegetarian burger with a meaty taste and texture. Available in natural food stores, usually in the freezer case.

bulgur—a hard red winter wheat that has been cracked and toasted. It cooks quickly and has a delicious, nutty flavor.

carob powder—the roasted powder of the carob bean, which can be used in place of chocolate in many recipes. Sold in natural food stores and some supermarkets.

couscous—called Middle Eastern pasta, it is made from the same type of wheat as pasta. Available in the grain section of many supermarkets as well as in natural food stores and ethnic markets.

The Menu Plan

date pieces—pitted, chopped dates that are coated with cornstarch to keep them from sticking together. Sold in natural food stores and some supermarkets.

diced green chilies—refers to diced Anaheim chilies, which are mildly hot. Available canned (Ortega is one brand) or fresh. When using fresh chilies, remove skin by charring it under a broiler and rubbing it off.

Emes Jel—a natural gelling agent made from vegetable ingredients. May be combined with fruit juice to make a natural "Jell-O." Also used to make Corn Butter (page 200). Sold in natural food stores.

Fat-free Nayonaise—a fat-free, cholesterol-free mayonnaise substitute that contains no dairy products or eggs. Sold in natural food stores.

Fines Herbs—an herb blend that usually features equal parts of tarragon, parsley, and chives, and may also contain chervil. Look for it in the spice section.

garlic granules—a granulated form of garlic powder that remains free-flowing.

Harvest Burger for Recipes—ready-to-use ground beef substitute made from soy. Ideal for tacos, pasta sauces, and chili. Made by Green Giant (Pillsbury) and available in supermarket frozen food sections.

instant bean flakes—precooked black or pinto beans that can be quickly reconstituted with boiling water and used as a side dish, dip, sauce, or burrito filling. Fantastic Foods and Taste Adventure are two brands. Available in natural food stores and some supermarkets.

Italian Seasoning—a mixture of Italian herbs: basil, oregano, thyme, marjoram, etc. May also be called Italian Herbs. Look for it in the spice section of your market.

jicama ("hick-ama")—a delicious root vegetable that is a crisp, slightly sweet addition to salads. Usually sold in the unrefrigerated area of your supermarket's produce section.

kudzu (or *kuzu*)—a thickener made from the roots of the kudzu vine, which grows rampantly in the southeastern United States. It is used much like arrowroot or cornstarch, and makes a creamy sauce or gel.

lite soy sauce—May also be called reduced-sodium soy sauce. Compare labels to find the brand with the lowest sodium content.

miso—a salty soybean paste used to flavor soup, sauces, and gravies. Available in several varieties, with the lighter-colored forms having the mildest flavor. Sold in natural food stores and Asian markets.

nori—a sea vegetable that comes in square sheets and is used for making nori rolls or sushi. Sold in natural food stores and Asian markets.

Pacific Cream Flavored Sauce Base—a nondairy cream soup base made by Pacific Foods of Oregon and sold in natural food stores and many supermarkets.

potato flour—used as a thickener in sauces, puddings, and gravies. One common brand sold in natural food stores and many supermarkets is Bob's Red Mill (see Product Guide/Resources).

prewashed salad mix, prewashed spinach—mixtures of lettuce, spinach, and other salad ingredients that have been cleaned and dried. These store well and make salad preparation a snap. Several different mixes are available in the produce department of most food stores. "Spring mix" is particularly flavorful.

prune puree—also called prune butter. Can be used in place of fat in baked goods. Commercial brands are WonderSlim and Lekvar. Prune baby food or pureed stewed prunes may also be used.

red pepper flakes—dried, crushed chili peppers. Available in the spice section or with the Mexican foods in your supermarket.

reduced-fat tofu—contains about one-third less fat than regular tofu. Three brands are MoriNu Lite, White Wave, and Tree of Life. Sold in natural food stores and supermarkets.

reduced-sodium soy sauce—See *lite soy sauce.*

Rice Dream—see *rice milk*

rice flour—a thickener for sauces, puddings, and gravies. Choose white rice flour for the smoothest results. Bob's Red Mill is one common brand sold in natural food stores and many supermarkets (see Product Guide/Resources).

rice milk—a beverage made from partially fermented rice that can be used in place of dairy milk on cereal and in most recipes. Available in natural food stores and some supermarkets.

roasted red peppers—roasted red bell peppers. These add great flavor and color to a variety of dishes. You can purchase them already roasted, packed in water, in most grocery stores. Usually located near the pickles.

seasoned rice vinegar—a mild vinegar, seasoned with sugar and salt. Great for salad dressings and on cooked vegetables. Available in most grocery stores with the vinegar or in the Asian foods section.

seitan ("say-tan")—also called wheat meat, seitan is a high-protein, fat-free food with a meaty texture and flavor. Available in the deli case or freezer of natural food stores.

silken tofu—a smooth, delicate tofu that is excellent for sauces, cream soups, and dips. Often available in special packaging that allows storage without refrigeration for up to a year. Refrigerate after opening. One popular brand, Mori-Nu, is available in most grocery stores. Ask for the reduced fat "Lite" variety.

sodium-free baking powder—made with potassium bicarbonate instead of sodium bicarbonate. Sold in natural food stores and some supermarkets. Featherweight is one brand.

soy milk—made from soybeans. Use as a beverage, on cereal, or for dairy milk and cream in most recipes. Available in regular, low-fat, fat-free, and calcium-fortified varieties. Available in natural food stores and many supermarkets.

soy milk powder—can be used in smoothies, baked goods, or mixed with water for a beverage. Available in natural food stores and some supermarkets or by mail order (see Product Guide/Resources).

Spike—a seasoning mixture of vegetables and herbs. Comes in a salt-free version as well as the original version, which contains salt. (The original version was used for the recipes in this book.) Sold in natural food stores and many supermarkets.

spreadable fruit—natural fruit preserves made of 100 percent fruit with no refined sugar.

stoneground mustard—mustard containing whole mustard seeds and often horseradish.

tahini—sesame seed butter. Comes in raw and toasted forms (either will work in these recipes, though the toasted form is preferred). Sold in natural food stores, ethnic grocery stores, and some supermarkets. Look for it near the peanut butter.

textured vegetable protein (TVP)—meatlike ingredient made from defatted soy flour. Rehydrate with boiling water to add protein and meaty texture to sauces, chili, and stews. TVP is sold in natural food stores and in the bulk section of some supermarkets.

unbleached flour—white flour that has not been chemically whitened. Available in most grocery stores.

turbinado sugar—unbleached sugar, available in natural food stores and most supermarkets.

vegetable broth—ready-to-use brands include Pacific Foods, Imagine Foods, and Swansons. Other brands available in dry form as powder or cubes. Sold in natural food stores and many grocery stores.

whole wheat pastry flour—milled from soft spring wheat, it retains the bran and germ, and at the same time produces lighter-textured baked goods than regular whole wheat flour. Available in natural food stores.

Yves Veggie Breakfast Links—fat-free vegetarian sausage. Sold in natural food stores and many supermarkets.

Yves Veggie Ground Round—fat-free ground beef substitute made from soybeans. Use in spaghetti sauce, tacos, and other hearty dishes. Sold in natural food stores and many supermarkets.

The Recipes

BREAKFAST FOODS

⁂ Multigrain Cereal ⁂

Makes about 2½ cups

A variety of multigrain hot cereal mixes are available in natural food stores and many supermarkets. One widely distributed brand is Bob's Red Mill 10 Grain Cereal; another is Arrowhead Mills Bear Mush Hot Breakfast Cereal. The method for cooking any of these hearty, satisfying breakfast foods is given below.

3 cups water
1 cup multigrain cereal mix
½ teaspoon salt (optional)*

Combine all ingredients in a saucepan and bring to a simmer. Cover loosely and simmer for 7 minutes. Remove from heat and let stand, covered, for 10 minutes before serving.

Per ⅔ cup: 180 calories; 6 g protein; 35 g carbohydrate; 2.5 g fat; 5–290 mg sodium

---

*HEALTH HINT: By gradually reducing the amount of salt you add, you can train your taste buds so that the cereal will taste fine with no salt at all.

❈ Rolled Grain Cereal ❈

Makes about 2 cups

Everyone is familiar with rolled oats, but few people are aware that several other whole grains are available in this easy-to-cook form. Check your natural food store or supermarket for rolled wheat, rye, barley, or triticale. Some are sold in boxes in the cereal section, but others may be found in the bulk food department. These can be cooked individually or mixed to make a multigrain cereal.

> 2 cups water
> 1 cup rolled grain
> ¼ teaspoon salt (optional)

Combine all ingredients in a saucepan and bring to a simmer. Cover loosely and simmer for 7 minutes. Remove from heat and let stand, covered, for 5 minutes before serving.

Per ½ cup: 79 calories; 2.5 g protein; 16 g carbohydrate; 0.4 g fat; 0–144 mg sodium

❈ Whole Wheat Cereal ❈

Makes about 2 cups

Whole wheat berries make a crunchy and satisfying breakfast cereal. Spring wheat, which is also called soft wheat or white wheat, is a variety that cooks in about 20 minutes if it is presoaked. Look for it in natural food stores or your supermarket's bulk section.

> 1 cup spring wheat berries

Rinse the wheat, then soak it overnight in 3 cups of water. In the morning, drain off the soaking water and put the wheat into a pan with ¾ cup of fresh water. Cover and simmer until tender, about 20 minutes.

VARIATION: For a different flavor, substitute whole oats or whole rye berries for the wheat berries.

Per ½ cup: 75 calories; 3 g protein; 16 g carbohydrate; 0 g fat; 2 mg sodium

Breakfast Teff

Makes 2 cups

Teff has been a staple grain in northern Africa for centuries. It has recently become available in the United States to the delight of chefs and nutritionists. This tiny grain is extremely nutritious and makes an absolutely delicious breakfast cereal with a wheatlike flavor. Ask for teff at your favorite natural food store.

½ cup teff
1½ cups water
¼ teaspoon salt (optional)

soy milk or rice milk for serving

Combine teff and water in a saucepan. Add salt if desired and stir to mix. Cook over low heat, stirring occasionally, until thick, about 15 minutes. Serve with soy milk or rice milk.

Per ½ cup: 100 calories; 4 g protein; 18 g carbohydrate; 1 g fat; 134 mg sodium; 0 mg cholesterol

❈ Muesli ❈

Makes about 5 cups

Muesli is a European breakfast cereal made of uncooked grains, dried fruits, and nuts. Serve it with hot or cold soy milk, rice milk, or fruit juice. The recipe makes about 5 cups.

> 3½ cups rolled oats
> ½ cup dried fruit (apples, figs, apricots, etc.), chopped
> ½ cup raisins
> ¼ cup slivered almonds

Combine all ingredients. They may be left whole or ground in a food processor until they are a fairly fine, uniform texture. Store in an airtight container in the refrigerator.

Per ½ cup: 180 calories; 6 g protein; 29 g carbohydrate; 4 g fat; 2 mg sodium

❈ Quick Breakfast Pudding ❈

Makes 3 cups

> 8–10 dried apricot halves
> 5–6 dried figs
> ¼ cup raisins
> 1 green apple (pippin or Granny Smith work well)
> 1 cup quick rolled oats
> 3 cups Vanilla Rice Dream Beverage
> ¼ teaspoon cinnamon

Chop the apricots, figs, and raisins in a food processor. Cut the apple and remove the core. Add the apple to the dried fruit in the food processor and chop finely. Transfer to a saucepan and add remaining ingredients. Simmer slowly over medium heat until thickened, about 5 minutes. Stir occasionally during cooking.

Per ½ cup: 160 calories; 5 g protein; 30 g carbohydrate; 2 g fat; 47 mg sodium

❈ Breakfast Sweet Potato Pudding ❈

Makes about 1½ cups

This breakfast will appeal to lovers of hot cereal and porridge. It takes just moments to prepare if you have cooked sweet potatoes or yams on hand.

⅓ cup rolled oats
½ cup soy milk or rice milk
1 cup cooked sweet potato or yam
1 tablespoon maple syrup
¼ teaspoon cinnamon

Combine all ingredients in a blender and blend until smooth.

Per ½ cup: 119 calories; 3 g protein; 24 g carbohydrate; 1 g fat; 21 mg sodium

❈ Breakfast Scramble ❈

Makes about 4 cups

This hearty breakfast is easy to make if you have leftover polenta.

2 teaspoons olive oil
½ cup chopped yellow onion

The Recipes

10 mushrooms, sliced

½ cup diced red or green bell pepper

1 teaspoon dried basil

½ teaspoon dried oregano

½ teaspoon dried thyme

½ teaspoon salt

¼ teaspoon black pepper

2–3 cups cooked, chilled polenta (page 273)

2–3 cups washed and chopped fresh spinach (optional)

Heat the oil in a large nonstick skillet and add the onion, mushrooms, and bell pepper. Cook over medium heat, stirring occasionally, for 3 minutes. Stir in the seasonings and continue cooking until the onion is soft and the mushrooms are browned. Add a tablespoon or two of water if needed to prevent sticking.

Cut the polenta into ½-inch cubes. Add to the skillet and use a spatula to gently fold the mixture together. If using the spinach, spread it evenly over the polenta mixture. Cover the pan and cook until the spinach is wilted and the mixture is completely hot, about 5 minutes. Toss gently to mix.

Per cup: 129 calories; 4 g protein; 22 g carbohydrate; 2 g fat; 336 mg sodium; 0 mg cholesterol

Whole Wheat Pancakes

Makes twenty-four 2-inch pancakes

Five simple ingredients are all it takes to make nutritious whole grain pancakes. Serve them with fresh fruit, unsweetened spreadable fruit, or maple syrup.

1 banana
1¼ cups soy milk or rice milk
1 tablespoon maple syrup
1 cup whole wheat pastry flour or whole wheat flour
2 teaspoons sodium-free baking powder
¼ teaspoon salt

fresh fruit, spreadable fruit, or maple syrup for serving

In a large bowl, mash the banana, then stir in the soy milk or rice milk and maple syrup. In a separate bowl, mix the flour, baking powder, and salt together; then stir them into the banana mixture.

Preheat a nonstick skillet or griddle, then spray it lightly with a vegetable oil spray. Add small amounts of the batter and cook until the tops bubble. Turn with a spatula and cook the flip sides until golden brown. Serve at once.

Per pancake: 29 calories; 1 g protein; 6 g carbohydrate; 0 g fat; 30 mg sodium

Multigrain Pancakes

Makes twenty 3-inch pancakes

1 cup soy milk or rice milk
2 tablespoons apple cider vinegar
1 tablespoon maple syrup
¼ cup Bob's Red Mill 10 Grain Cereal or similar multigrain cereal
½ cup whole wheat pastry flour
½ teaspoon baking soda
¼ teaspoon salt

fresh fruit, spreadable fruit, or maple syrup for serving

Mix the milk, vinegar, maple syrup, and cereal and let stand 10 minutes. In a separate bowl, combine the flour, soda, and salt. Add to the milk mixture and stir to mix.

Preheat a nonstick skillet and mist lightly with vegetable oil spray. Pour in enough batter to make several small pancakes. Cook until the tops bubble and the edges are dry, then turn carefully with a spatula and cook the flip sides until golden brown. Serve at once.

Per pancake: 26 calories; 1 g protein; 5 g carbohydrate; 0 g fat; 54 mg sodium

Sourdough Waffles

Makes about six 7-inch waffles

These waffles are light and crisp with a tangy flavor.

2 cups sourdough starter
½ cup cornmeal

1 teaspoon baking soda

2 tablespoons turbinado sugar, maple syrup, or
other sweetener

Mix all of the ingredients together, preferably in a large measuring cup so the batter can be easily poured onto the waffle iron. Preheat the waffle iron, then lightly coat it with an oil spray. Add just enough batter to make a thin layer that reaches to the edges of the waffle iron. Cook, without lifting the lid, until the waffle is crisp and golden brown, 3 to 5 minutes. Serve immediately with fresh fruit, maple syrup, or spreadable fruit.

TIP: This batter can also be used to make pancakes.

Per waffle: 136 calories; 3 g protein; 30 g carbohydrate; 0 g fat; 138 mg sodium

✖ Buckwheat Bananacakes ✖

Makes twenty-four 2-inch pancakes

If you like the distinctive flavor of buckwheat, you'll love these easily prepared pancakes. Enjoy them with applesauce, fresh fruit, or maple syrup.

½ cup buckwheat flour

½ teaspoon sodium-free baking powder

¼ teaspoon baking soda

¼ teaspoon salt

½ cup rolled oats

1 ripe banana

2 tablespoons maple syrup

1 tablespoon vinegar

1 cup soy milk or rice milk

The Recipes

fresh fruit, spreadable fruit, or maple syrup
for serving

Stir the buckwheat flour, baking powder, baking soda, and salt together in a mixing bowl. Place the rolled oats into a blender with the banana, maple syrup, vinegar, and soy or rice milk. Blend until smooth, then add to the flour mixture, stirring to remove any lumps.

Preheat a nonstick skillet or griddle, then lightly spray it with vegetable oil spray. Pour small amounts of batter onto the heated surface and cook over medium-high heat until the tops bubble. When the bubbles break and the edges of the pancakes are dry, turn them carefully with a spatula and cook the second sides until golden brown, 30 to 60 seconds. Serve immediately.

Per pancake: 27 calories; 1 g protein; 5.5 g carbohydrate; 0 g fat; 21 mg sodium

✸ Cornmeal Flapjacks ✸

Makes about sixteen 2-inch pancakes

½ cup cornmeal
½ cup whole wheat pastry flour
¼ teaspoon sodium-free baking powder
¼ teaspoon baking soda
¼ teaspoon salt
1 cup soy milk or rice milk
2 tablespoons maple syrup
1 tablespoon vinegar

fresh fruit, spreadable fruit, or maple syrup for
serving

Turn Off the Fat Genes

Stir the cornmeal, whole wheat flour, baking powder, baking soda, and salt together in a mixing bowl. In a separate bowl, combine the milk, syrup, and vinegar. Add the flour mixture and stir just enough to make a smooth batter.

Preheat a nonstick skillet or griddle, then mist it with vegetable oil spray. Pour small amounts of batter onto the heated surface and cook over medium-high heat until the tops are fairly dry. Turn carefully with a spatula and cook the second sides until golden brown, about 30 seconds. Serve immediately with fresh fruit, spreadable fruit, or maple syrup.

Per pancake: 33 calories; 1 g protein; 7 g carbohydrate; 0 g fat; 42 mg sodium

Oatmeal Waffles

Makes 6 waffles

Oatmeal waffles are my own personal favorite. They are substantial and chewy, a bit like warm, hearty oatmeal with a crunchy crust.

- 2 cups rolled oats
- 2 cups water
- 1 banana
- 1 teaspoon sodium-free baking powder
- ¼ teaspoon salt
- 1 tablespoon maple syrup
- 1 teaspoon vanilla

fresh fruit, spreadable fruit, or maple syrup for serving

Place all the ingredients into a blender and blend until smooth. Cook in a preheated, oil-sprayed waffle iron for 10 minutes without lifting the lid. The cooking time may vary slightly with different waffle irons. Serve with fresh fruit.

NOTE: The batter should be pourable. If it becomes too thick as it stands, add a bit more water to achieve the desired consistency.

Per waffle: 138 calories; 5 g protein; 25 g carbohydrate; 2 g fat; 89 mg sodium

Sweet Potato Waffles

Makes two 10-inch waffles

These delicious golden waffles will be a hit with everyone.

2 cups rolled oats
2½ cups water
1 cup cooked sweet potato or yam (about 2 medium sweet potatoes)
1 ripe banana
2 tablespoons maple syrup
½ teaspoon sodium-free baking powder
¼ teaspoon cinnamon
¼ teaspoon salt
1 teaspoon vanilla

Place all the ingredients into a blender and blend until completely smooth. Preheat a waffle iron and spray it with vegetable oil spray. Spread with a layer of batter, then cook for 8 to 10 minutes without lifting the lid. The cooking time may vary slightly with different waffle irons. Serve with maple syrup.

NOTE: The batter should be pourable. If it becomes too thick as it stands, add a bit more water to achieve the desired consistency.

Per waffle: 203 calories; 8 g protein; 33 g carbohydrate; 3 g fat; 134 mg sodium

❋ French Toast ❋

Makes 6 slices

1 cup soy milk
¼ cup whole wheat pastry flour
1 tablespoon maple syrup
1 teaspoon nutritional yeast (optional)
1 teaspoon vanilla
½ teaspoon cinnamon
6 slices whole wheat bread

fresh fruit, spreadable fruit, or maple syrup for
    serving

Combine and whisk or blend all ingredients except the bread until smooth. Pour into a flat dish. Soak the bread in batter until it is soft but not soggy. Cook in an oil-sprayed nonstick skillet until golden brown, 2 to 3 minutes per side.

Per slice: 111 calories; 4 g protein; 21 g carbohydrate; 1 g fat; 116 mg sodium

BREADS AND BAKED GOODS

❋ Quick and Easy Brown Bread ❋

Makes 1 loaf (about 20 slices)

This bread is similar to Boston brown bread, sweet and moist with no added fat or oil. It is quick to mix and requires no kneading or rising.

1½ cups soy milk
2 tablespoons vinegar
3 cups whole wheat pastry flour
2 teaspoons baking soda
½ teaspoon salt

½ cup molasses
½ cup raisins

orange marmalade for serving (optional)

Preheat oven to 325° F. Mix the soy milk with vinegar and set aside. In a large bowl, stir the whole wheat pastry flour, soda, and salt together. Add the soy milk mixture and molasses. Stir to mix, then stir in the raisins. Do not overmix. Spoon into a large nonstick or oil-sprayed loaf pan and bake for 1 hour.

Per slice: 100 calories; 3 g protein; 21 g carbohydrate; 0 g fat; 149 mg sodium

❋ Corn Bread ❋

Serves 9

This corn bread is delicious and crumbly with no added fat. The secret is barley flour, which can be found in natural food stores and many supermarkets (see Product Guide/Resources).

1½ cups soy milk
1½ tablespoons vinegar
1 cup cornmeal
1 cup barley flour
2 tablespoons sugar or other sweetener
1 tablespoon baking powder
½ teaspoon baking soda
½ teaspoon salt

Preheat oven to 425° F. Combine the soy milk and vinegar and set aside. Mix the cornmeal, barley flour, sugar, baking powder, baking soda, and salt in a large bowl. Add the soy milk mixture and stir until just mixed. Spread the batter evenly in an oil-sprayed 9 × 9-inch baking dish. Bake until the top is golden brown, 25 to 30 minutes. Serve hot.

Per serving: 125 calories; 3 g protein; 26 g carbohydrate; 1 g fat; 180 mg sodium

❈ Garlic Bread I (mild) ❈

Makes 1 loaf (20 slices)

1 cup Corn Butter (page 200)
2 teaspoons garlic granules or powder (or more to taste)
1 teaspoon Italian Seasoning
1 loaf whole wheat French bread

Preheat oven to 350° F. Mix the Corn Butter, garlic granules, and Italian Seasoning together until smooth. Cut the bread and spread with the mixture. Wrap in foil and bake 30 minutes.

Per ½-inch slice: 76 calories; 3 g protein; 14 g carbohydrate; 1 g fat; 139 mg sodium

❈ Garlic Bread II (strong) ❈

Makes 1 loaf (20 slices)

1 cup Corn Butter (page 200)
1–2 tablespoons finely minced fresh garlic
1 teaspoon Italian Seasoning
1 loaf whole wheat French bread

Preheat oven to 350°F. Mix the Corn Butter, minced garlic, and Italian Seasoning together until smooth. Cut the bread and spread with the mixture. Wrap in foil and bake 30 minutes.

Per ½-inch slice: 78 calories; 3 g protein; 14 g carbohydrate; 1 g fat; 140 mg sodium

❈ Double Bran Muffins ❈

Makes 12 muffins

These wholesome, fruity muffins contain two types of bran. Prune Puree makes them moist without added fat. Make your own (page 206) or look for it under the names WonderSlim or Lekvar, or use a 4-ounce jar of prune baby food. The muffins will be quite moist when they first come out of the oven, so let them stand a few minutes before serving.

2 cups whole wheat or whole wheat pastry
  flour
¾ cup wheat bran
¾ cup oat bran
½ teaspoon salt
1 teaspoon baking soda
1 teaspoon cinnamon
¼ teaspoon nutmeg
1 apple, finely chopped or grated (use a food
  processor)
½ cup raisins
1½ cups soy milk
1½ tablespoons vinegar
¼ cup Prune Puree (page 206) or prune baby
  food
⅓ cup molasses

Turn Off the Fat Genes

Preheat oven to 350° F. Mix together the flour, brans, salt, soda, and spices. In a separate bowl, mix together the remaining ingredients. Combine the wet and dry ingredients and stir to mix. Spoon into muffin pans that have been sprayed with a nonstick spray and bake until the tops bounce back when lightly pressed, about 25 minutes. Let stand 1 to 2 minutes, then remove from pan and let stand 5 minutes before serving.

Per muffin: 169 calories; 5 g protein; 35 g carbohydrate; 1 g fat; 178 mg sodium

Fresh Apple Muffins

Makes 12 muffins

These muffins make a delicious and satisfying breakfast or snack.

3 cups whole wheat pastry flour
½ cup sugar or other sweetener
2 teaspoons baking soda
½ teaspoon salt
2 cups finely chopped green apple (about 2 medium-large apples)
1½ cups soy milk or rice milk
2 tablespoons cider vinegar
⅓ cup blackstrap molasses*
¾ cup raisins
¼ cup chopped walnuts

* Blackstrap molasses, which is a rich source of calcium and iron, is not particularly sweet—hence the sugar in the recipe. If you choose to substitute regular molasses, which is significantly sweeter, you can reduce the amount of sugar, or eliminate it altogether, depending on how sweet you like your muffins.

Preheat oven to 375° F. Mix together the flour, sugar, baking soda, and salt in a mixing bowl.

Quarter and core the apples, then chop them finely in a food processor or by hand. Place them into a large mixing bowl and add the milk, vinegar, and molasses. Stir until just mixed, then stir in the raisins and walnuts.

Lightly spray muffin cups with vegetable oil spray and fill to the top with batter. Bake 30 to 35 minutes, until the tops bounce back when lightly pressed. Remove the muffins from the oven and let stand about 5 minutes, then remove from the pan and cool on a rack. Store in an airtight container.

Per muffin: 219 calories; 5 g protein; 45 g carbohydrate; 2 g fat; 249 mg sodium

✳ Sweet Potato Muffins ✳

Makes 10 to 12 muffins

2 cups whole wheat or whole wheat pastry flour
½ cup sugar
1 tablespoon baking powder
½ teaspoon baking soda
½ teaspoon salt
½ teaspoon cinnamon
¼ teaspoon nutmeg
1¾ cups cooked sweet potatoes, or 1 15-ounce can, drained
½ cup water
½ cup raisins

Preheat oven to 375° F. Mix together the flour, sugar, baking powder, baking soda, salt, cinnamon, and nutmeg in a large bowl. Mash the sweet potatoes and add them along with the water and raisins. Stir until just mixed.

Turn Off the Fat Genes

Lightly oil-spray muffin cups and fill to the top. Bake 25 to 30 minutes, until the tops of the muffins bounce back when pressed lightly. Let stand 1 to 2 minutes before removing from the pan. When cool, store in an airtight container in the refrigerator.

Per muffin: 148 calories; 3 g protein; 34 g carbohydrate; 0 g fat; 128 mg sodium

DIPS, SAUCES, AND DRESSINGS

Red Pepper Hummus

Makes about 2 cups

Serve this delicious dip with fresh vegetables or pita wedges or use it as a spread on sandwiches.

1 15-ounce can garbanzo beans, drained
½ cup water-packed roasted red pepper (about 2 peppers)
1 tablespoon tahini (sesame seed butter)
3 tablespoons lemon juice
1 garlic clove (or more to taste)
¼ teaspoon ground cumin

fresh vegetables or pita bread for serving

Combine all ingredients in a food processor and process until completely smooth.

Per tablespoon: 19 calories; 1 g protein; 3 g carbohydrate; 0.4 g fat; 19 mg sodium

Quick Bean Dip

Makes about 2 cups

Dehydrated bean flakes can be used to make a quick and tasty dip for baked tortilla chips. They are sold in packages as well as in bulk in many natural food stores and supermarkets. Fantastic Foods and Taste Adventure are two common brands.

1 cup instant bean flakes
1 cup boiling water
1 cup salsa (you choose the heat)

Mix the bean flakes and boiling water and let stand for 5 minutes. Stir in salsa.

Per ¼ cup: 64 calories; 4 g protein; 12 g carbohydrate; 0 g fat; 77 mg sodium

Zuccamole

Makes about 1 cup

Serve this fat-free guacamole alternative as a topping for tostadas or burritos, or as a dip with baked tortilla chips.

1½ cups sliced zucchini
2 tablespoons chopped green onions
1 garlic clove, peeled
2 teaspoons lemon juice
¼ teaspoon cumin
¼ teaspoon salt
½ cup diced tomato

Steam the zucchini until it is tender-crisp. Remove from steamer and cool. Puree the cooled zucchini in a food processor for 2 to 3 minutes, until very smooth. Add the green onions, garlic, lemon juice, cumin, and salt and process briefly. Transfer from the processor to a serving bowl and stir in the diced tomato.

Per tablespoon: 4 calories; 0 g protein; 1 g carbohydrate; 0 g fat; 37 mg sodium

❋ Vicki's Tofu Mayo ❋

Makes about 1 cup

Vicki Saunders, the dietitian for the McDougall Program at St. Helena Health Center, developed this recipe using silken tofu as the main ingredient. One brand of silken tofu, Mori-Nu, is sold in most supermarkets.

> 1 8-ounce package reduced-fat firm silken tofu
> ½ teaspoon sugar
> ½ teaspoon salt
> ½ teaspoon prepared mustard
> 2 teaspoons lemon juice
> 2 teaspoons vinegar

Combine all ingredients in a food processor or blender. Blend until completely smooth. This will take several minutes. Chill.

Per tablespoon: 15 calories; 0.2 g protein; 1 g carbohydrate; 0.4 g fat; 110 mg sodium

Corn Butter

Makes about 2 cups

This golden spread may be used in place of butter on bread, vegetables, and many other foods. It is made with Emes Jel, a completely vegan product sold in natural food stores (see Product Guide/Resources, page 318).

- 2 teaspoons Emes Jel, or 1½ teaspoons agar powder
- ¼ cup cold water
- 1 cup boiling water
- 1 cup cornmeal mush*
- 2 tablespoons raw cashews
- ½ teaspoon salt
- 2 teaspoons lemon juice
- 1 tablespoon finely grated raw carrot
- 1 teaspoon nutritional yeast (optional)

Place the Emes Jel or agar powder in a blender with the cold water and let stand 3 to 5 minutes. Add 1 cup boiling water and blend to dissolve. Add remaining ingredients and process until *totally smooth* (this is essential and will take several minutes). Pour into a serving jar or dish and cover. Chill.

Per tablespoon (with cashews): 7 calories; 0.2 g protein; 1 g carbohydrate; 0.3 g fat; 36 mg sodium

---

*To make cornmeal mush, mix ¼ cup cornmeal with 1 cup water in a saucepan. Bring to a simmer and cook, stirring frequently, until thickened.

White Sauce

Makes 2 cups

Use this sauce to make mashed potatoes or creamed spinach or any other recipe requiring a white sauce.

2 cups soy milk
½ teaspoon onion powder
¼ teaspoon garlic powder
⅛ teaspoon salt
3 tablespoons rice flour
¼ cup potato flour
1 cup Corn Butter (page 200)

Pour the soy milk into a blender. With the blender running, add the ingredients in the order listed, allowing the blender to run long enough to make the mixture totally smooth.

Per ¼ cup: 75 calories; 2 g protein; 12 g carbohydrate; 2 g fat; 124 mg sodium

Brown Gravy

Makes about 2 cups

2 cups water or vegetable stock
1 tablespoon cashews
1 tablespoon onion powder
½ teaspoon garlic granules or powder
2 tablespoons cornstarch
½ teaspoon thyme
¼ teaspoon black pepper
2–3 tablespoons reduced-sodium soy sauce

Combine all ingredients in a blender and process until smooth, 2 to 3 minutes. Transfer to a saucepan and cook over medium heat, stirring constantly, until thickened.

Per ¼ cup: 18 calories; 1 g protein; 3 g carbohydrate; 0.5 g fat; 19 mg sodium

❋ Simple Marinara ❋

Makes about 2 cups

Serve this basic sauce with pasta or polenta, and top it with grilled or steamed vegetables.

> ½ cup red wine or water
> 1 small onion, chopped
> 2 garlic cloves, crushed
> 1 15-ounce can crushed or ground tomatoes
> 2 teaspoons mixed Italian herbs
> 1 tablespoon apple juice concentrate
> ¼ teaspoon black pepper

Heat the wine or water in a large pot, then add the onion and garlic and cook until soft, about 5 minutes. Add the tomatoes, Italian herbs, apple juice concentrate, and black pepper. Cover and simmer 15 minutes.

Per ½ cup: 50 calories; 2 g protein; 8 g carbohydrate; 0 g fat; 33 mg sodium; 0 mg cholesterol

Black Bean Sauce

Makes 2 cups

One of my favorite simple meals is a steamed red potato split open and stuffed with steamed broccoli and topped with this Black Bean Sauce.

1 15-ounce can black beans, drained
½ cup roasted red pepper
2 tablespoons lemon juice
2 tablespoons tahini (sesame seed butter)
½ teaspoon chili powder
¼ teaspoon cumin
¼ teaspoon coriander
¼ cup chopped fresh cilantro

Puree all of the ingredients in a food processor or blender until very smooth.

Per ¼ cup: 58 calories; 3 g protein; 9 g carbohydrate; 1 g fat; 60 mg sodium

Strawberry Sauce

Makes about 3 cups

1 cup water
1 cup apple juice concentrate
1½ teaspoons agar powder*
2 cups strawberries, fresh or frozen
¼ teaspoon cinnamon

---

*Agar is a thickener made from a sea vegetable that is an excellent replacement for animal gelatin. It is sold in natural food stores and Asian markets. For more information about agar, see Foods That May Be New to You, page 173.

Combine the water, apple juice concentrate, and agar powder in a saucepan and let stand 5 minutes. Blend 1 cup of the strawberries in a food processor or blender until fairly smooth and add to the apple juice. Bring to a simmer, stirring occasionally, and cook 3 minutes. Coarsely chop the remaining berries and add them, with the cinnamon, to the sauce. Stir to mix. Chill.

Per tablespoon: 12 calories; 0 g protein; 3 g carbohydrate; 0 g fat; 12 mg sodium

❉ Pineapple Apricot Sauce ❉

Makes about 3 cups

1 cup water
1 cup apple juice concentrate
1½ teaspoons agar powder
1 cup apricots, fresh, frozen, or canned
1 cup crushed pineapple
¼ teaspoon ginger

Combine the water, apple juice concentrate, and agar powder in a saucepan and let stand 5 minutes. Bring to a simmer, stirring occasionally, and cook 3 minutes. Coarsely chop the apricots and add them, along with the pineapple and ginger. Stir to mix. Chill.

Per tablespoon: 14 calories; 0.3 g protein; 3 g carbohydrate; 0 g fat; 12 mg sodium

Turn Off the Fat Genes

✳ Date Sauce ✳

Serves 9

Use as a syrup on pancakes.

2 cups chopped pitted dates or date pieces
3 cups water
1 teaspoon vanilla
⅛ teaspoon salt

Combine the date pieces and water in a saucepan and bring to a simmer. Cook until the date pieces are soft, stirring frequently, then transfer to a blender. Add the remaining ingredients and blend until smooth.

NOTE: When blending hot ingredients, be sure to start the blender on low speed and hold the lid on tightly.

Per ¼ cup: 61 calories; 0 g protein; 15 g carbohydrate; 0 g fat; 19 mg sodium

✳ Plum Sauce ✳

Makes about 2 cups

1 17-ounce can purple plums in heavy syrup
2 garlic cloves
1 tablespoon cornstarch
2 tablespoons seasoned rice vinegar
1 tablespoon soy sauce
⅛ teaspoon cayenne (more or less to taste)

Remove pits from the plums, then puree plums in a blender or food processor along with their liquid and the remaining ingredients. Heat in a saucepan, stirring constantly, until thickened.

Per tablespoon: 17 calories; 0 g protein; 2 g carbohydrate; 0 g fat; 40 mg sodium

The Recipes

✼ Cranberry Apple Sauce ✼

Makes 6 cups

2 cups unsweetened applesauce
2 12-ounce packages fresh cranberries
1 6-ounce can apple juice concentrate
1 6-ounce can orange juice concentrate
1 tablespoon agar flakes
1 cup sugar or other sweetener
¾ teaspoon cinnamon

Combine all ingredients in a pan and simmer, uncovered, until the cranberries are completely soft, about 20 minutes.

Per serving: 128 calories; 0 g protein; 32 g carbohydrate; 0 g fat; 14 mg sodium; 0 mg cholesterol

✼ Prune Puree ✼

Makes 1 cup

Prune Puree serves as a substitute for fat and sugar in many baking recipes.

1 cup pitted prunes

Place the prunes in a small saucepan with just enough water to cover. Cover the pan, bring to a simmer, and cook until the prunes are soft, about 15 minutes. Drain and cool.

Puree the cooled prunes in a food processor fitted with the steel blade.

Per tablespoon: 24 calories; 0 g protein; 6 g carbohydrate; 0 g fat; 0 mg sodium

Apple Chutney

Makes about 1 quart

Chutney is a spicy relish that is served as a condiment with Indian meals. This simple chutney is made with apples. It keeps in the refrigerator for several weeks and also freezes for longer storage.

- 3 large tart green apples (about 1½ pounds)
- 1 cup cider vinegar
- 1 cup sugar or other sweetener
- 1 large garlic clove, minced
- 1 tablespoon minced ginger root, or ½ teaspoon ginger powder
- ½ cup orange juice
- 1 teaspoon each: cinnamon, cloves
- ½ teaspoon salt
- ¼ teaspoon cayenne (or more to taste)

Core and coarsely chop the apples. Combine them with all the remaining ingredients in a saucepan. Bring to a simmer and cook uncovered, stirring occasionally, until most of the liquid is absorbed, about 1 hour.

Per tablespoon: 22 calories; 0 g protein; 5 g carbohydrate; 0 g fat; 21 mg sodium; 0 mg cholesterol

Sesame Salt

Makes ½ cup

Sesame salt is a delicious seasoning for steamed vegetables, cooked grains, and legumes.

½ cup unhulled sesame seeds
½ teaspoon salt

Toast the sesame seeds in a dry skillet over medium heat, stirring constantly until the seeds begin to pop and brown slightly, about 5 minutes. Transfer to a blender, add the salt, and grind into a fine powder.

Per tablespoon: 54 calories; 1.5 g protein; 2.5 g carbohydrate; 4 g fat; 134 mg sodium; 0 mg cholesterol

Sesame Seasoning

Makes ½ cup

This seasoning adds delicious flavor to steamed vegetables, cooked rice, or baked potatoes. Unhulled sesame seeds (sometimes called brown sesame seeds) are sold in natural food stores and specialty shops.

½ cup unhulled sesame seeds
2 tablespoons nutritional yeast flakes
½ teaspoon salt

Toast the sesame seeds in a dry skillet over medium heat, stirring constantly until the seeds begin to pop and brown slightly, about 5 minutes. Transfer to a blender along with the nutritional yeast and salt and grind into a powder.

Per tablespoon: 58 calories; 2 g protein; 3 g carbohydrate; 4 g fat; 137 mg sodium

❈ Cucumbers with Creamy ❈ Dill Dressing

Makes about 4 cups

This salad features cool cucumber slices in a creamy tofu dressing. Use reduced-fat tofu.

> 8 ounces firm reduced-fat tofu
> 1 teaspoon garlic granules or powder
> ½ teaspoon dill weed
> ¼ teaspoon salt
> 2 tablespoons lemon juice
> 3 tablespoons seasoned rice vinegar
> 1 tablespoon cider vinegar
> 2 cucumbers, peeled and thinly sliced
> ¼ cup thinly sliced red onion

Blend the tofu, garlic granules, dill weed, salt, lemon juice, and vinegars in a food processor for 2 to 3 minutes, until completely smooth. Pour dressing over cucumbers and onion and toss to mix. Chill before serving.

Per ½ cup: 42 calories; 3 g protein; 5 g carbohydrate; 1 g fat; 181 mg sodium

❋ Texas Caviar ❋

Makes 3 cups

Serve with wedges of pita bread or crusty French bread.

1 15-ounce can black-eyed peas, drained
1 red or yellow bell pepper, seeded and diced
½ cup chopped green onions
1 tomato, diced
¼ cup chopped fresh cilantro (optional)
1 teaspoon finely chopped jalapeño pepper
¼ cup seasoned rice vinegar
1 teaspoon ground cumin

Bread for serving

Combine all ingredients in a large bowl and toss gently to mix. Chill 1 to 2 hours.

Per ½ cup: 74 calories; 4 g protein; 14 g carbohydrate; 0 g fat; 153 mg sodium

❋ Fresh Broccoli Salad ❋

Makes about 8 cups

1 bunch broccoli
1 large cucumber, cut in thick slices
½ cup finely sliced red onion
1–2 garlic cloves, minced
1 cup seasoned rice vinegar
1 tablespoon toasted sesame oil
½ teaspoon dried red pepper flakes

Cut broccoli into bite-sized florets, peel stems and cut into bite-sized pieces, and place in a salad bowl. Add remaining ingredients and toss to mix. Let stand 20 minutes before serving.

Per ½ cup: 44 calories; 1 g protein; 6 g carbohydrate; 1.5 g fat; 310 mg sodium

Three Bean Salad

Makes about 6 cups

This salad is delicious all by itself or as an addition to a green salad. I like to mix it with torn romaine lettuce leaves for a nearly instant salad.

1 15-ounce can kidney beans, drained
1 15-ounce can garbanzo beans, drained
1 15-ounce can green beans, drained
½ small red onion, finely chopped (optional)
¼ cup finely chopped fresh parsley
½ cup cider vinegar
¼ cup seasoned rice vinegar
3 garlic cloves, minced
2 tablespoons chopped fresh basil, or ½ teaspoon dried basil
½ teaspoon each: oregano, marjoram
¼ teaspoon black pepper

Place the drained beans in a large bowl with the chopped onion and parsley. In a separate bowl, whisk the vinegars, garlic, and seasonings together. Add to the beans and toss to mix. If possible, refrigerate for 2 to 3 hours before serving.

Per ½ cup: 85 calories; 4 g protein; 16 g carbohydrate; 0 g fat; 238 mg sodium

❋ Spinach Salad with Curry Dressing ❋

Serves 8

This wonderful spinach salad is a happy marriage of flavors and textures. It is especially easy to make when you use prewashed fresh spinach, available in the produce department of most markets.

½ bag prewashed spinach (about 6 cups loosely
   packed spinach leaves)
1 green apple, cored and diced
2 green onions, finely sliced, including green
   tops
¼ cup chopped dried apricots
3 tablespoons seasoned rice vinegar
3 tablespoons frozen apple juice concentrate
2 teaspoons stone-ground mustard
1 teaspoon reduced-sodium soy sauce
½ teaspoon curry powder
¼ teaspoon black pepper

Combine the washed spinach with the apple, onions, and apricots. Mix the vinegar, apple juice concentrate, mustard, soy sauce, curry powder, and black pepper in a small bowl and whisk together. Pour over salad and toss to mix just before serving.

Per serving: 45 calories; 1 g protein; 9 g carbohydrate; 0 g fat; 175 mg sodium

⁂ Antipasto Salad ⁂

Serves 8

The vegetables in this salad are steamed until just tender, then marinated in a vinaigrette dressing. This salad is good hot or cold.

2 large red potatoes, scrubbed
2 cups sliced carrots
2 cups frozen Italian green beans
2 cups cauliflower florets
1 red bell pepper, sliced or diced
¼ cup finely chopped parsley
¼ cup balsamic vinegar
2 tablespoons lemon juice
2 tablespoons seasoned rice vinegar
1 tablespoon apple juice concentrate
2 garlic cloves, crushed
2 teaspoons stone-ground or Dijon-style mustard
½ teaspoon salt
¼ teaspoon black pepper

Cut the potatoes into wedges and steam with the carrots over boiling water until just tender, about 10 minutes. Place in a salad bowl. Steam the green beans and cauliflower until just tender, 7 to 8 minutes. Add to the salad bowl along with the bell pepper and parsley. Mix the remaining ingredients together, pour over vegetables, and toss to mix. Serve warm or chilled.

Per serving: 74 calories; 2 g protein; 16 g carbohydrate; 0 g fat; 254 mg sodium

Ensalada de Frijoles

Serves 4 as a complete meal

This salad makes a perfect meal for hot summer days. It is quick to prepare, especially if you use prewashed salad mix. Jicama (pronounced "hick-ama") is a crisp root vegetable that is usually sold in the unrefrigerated area of your supermarket's produce section.

- 3 cups (approximately) cooked brown rice, or 4 warm corn tortillas
- 6 cups prewashed salad mix
- 1 carrot, grated or cut into small julienne strips
- 1 15-ounce can black beans, drained and rinsed
- 1 cup peeled and grated jicama
- 2 tomatoes, diced or cut into wedges
- 1 15-ounce can corn, drained, or 2 cups fresh or frozen
- ½ cup cilantro leaves, coarsely chopped (optional)
- ¼ cup salsa (your favorite variety)
- ¼ cup seasoned rice vinegar
- 1 garlic clove, crushed or pressed
- Additional salsa for topping

Warm the brown rice or corn tortillas in a microwave or on the stove. Divide among four plates, then cover with a layer of salad mix. Top with carrot, black beans, jicama, tomatoes, corn, and cilantro leaves. Mix together the salsa, seasoned rice vinegar, and crushed garlic. Sprinkle the salsa-vinegar dressing over each of the salads, then top with additional generous spoonfuls of salsa.

Per serving: 282 calories; 10 g protein; 56 g carbohydrate; 1.5 g fat; 324 mg sodium

❋ Rainbow Salad ❋

Makes 4 to 5 cups

This colorful salad is a delicious way to add cabbage to your diet.

3 cups shredded green cabbage
1½ cups shredded red cabbage
2 carrots, grated or julienned
2 stalks celery, thinly sliced
3 green onions, sliced
1 small green apple, finely chopped or julienned
1 tablespoon lemon juice
⅓ cup Vicki's Tofu Mayo (page 199) or Fat-free Nayonaise
⅓ cup apple juice concentrate

Place the red and green cabbage, the carrots, celery, and green onions in a salad bowl. Mix together the apple with the lemon juice and add it to the salad bowl. Add the Tofu Mayo and apple juice concentrate and mix well. Chill before serving.

Per ½ cup: 38 calories; 0.5 g protein; 5 g carbohydrate; 1.4 g fat; 71 mg sodium

❋ Potato Salad ❋

Serves 8

3 cups diced potatoes (preferably red or gold)
½ cup sliced celery, including leaves
½ cup finely chopped red onion
¼ cup finely chopped parsley
¼ cup cider vinegar

½ 16-ounce package reduced-fat firm tofu
¼ cup seasoned rice vinegar
1½ tablespoons stone-ground mustard
¼ teaspoon dill weed
½ teaspoon salt
⅛ teaspoon black pepper
⅛ teaspoon turmeric

Scrub the potatoes, peel if desired, and cut them into cubes. Steam over boiling water until just tender, about 15 minutes, then transfer them to a large bowl. Add the celery, onion, parsley, and cider vinegar. Stir to mix. Combine the remaining ingredients in a food processor or blender; blend until completely smooth, 2 to 3 minutes. Pour over salad and toss gently. Chill if time allows.

Per serving: 123 calories; 4.5 g protein; 24 g carbohydrate; 1 g fat; 328 mg sodium

✖ Tabouli ✖

Makes about 5 cups

Tabouli is a traditional Middle Eastern salad made with bulgur and seasoned with lemon, parsley, mint, and garlic. I know this will sound like heresy, but I never particularly liked tabouli until I left the mint out by accident. Then it tasted great! So now I always make it without mint. I include the mint in this recipe as an optional ingredient for those who enjoy it.

1 cup uncooked bulgur
2 cups boiling water
2 medium tomatoes, diced
½ cup chopped green onions, including green tops

Turn Off the Fat Genes

½ cucumber, peeled and diced

½ cup finely chopped parsley

3 tablespoons fresh mint leaves, finely chopped
  (optional)

¼ cup lemon juice

1 garlic clove, pressed or crushed

½–1 teaspoon salt

Put the bulgur in a large bowl and pour the boiling water over it. Cover and let stand until tender, about 25 minutes. Drain off any excess liquid and use a fork to fluff the bulgur. Add the tomatoes, green onions, cucumber, parsley, mint (if you like it!), lemon juice, garlic, and salt. Toss gently to mix. Chill before serving if time allows.

Per ½ cup: 78 calories; 3 g protein; 16 g carbohydrate; 0 g fat; 112–219 mg sodium

❋ Couscous Confetti Salad ❋

Makes about 6 cups

Couscous has its origins in northern Africa and is the world's smallest pasta. It cooks almost instantly and makes a beautiful and flavorful salad. Whole wheat couscous is sold in natural food stores and some supermarkets.

1½ cups whole wheat couscous

2 cups boiling water

3–4 green onions, finely chopped, including tops

1 red bell pepper, finely diced

1 carrot, grated

1–2 cups finely shredded red cabbage

½ cup finely chopped parsley

The Recipes

½ cup golden raisins or chopped dried apricots
1 lemon, juiced (about ¼ cup)
¼ cup seasoned rice vinegar
1 tablespoon olive oil
1 teaspoon curry powder
½ teaspoon salt

Place the couscous in a large bowl. Stir in the boiling water, then cover and let stand until all the water has been absorbed, 5 to 10 minutes. Fluff with a fork. Prepare all the vegetables as directed, then add them along with the remaining ingredients to the couscous. Stir to mix. Serve at room temperature or chilled.

Per ½ cup: 86 calories; 2 g protein; 17 g carbohydrate; 1 g fat; 107 mg sodium

✻ Sweet and Sour Cucumbers ✻

Makes 4 cups

4 cups peeled, sliced cucumber
½ cup thinly sliced red onion
3 tablespoons seasoned rice vinegar
1 teaspoon dried mint
¼ teaspoon black pepper

Mix all ingredients together; chill before serving.

Per ½ cup: 16 calories; 0 g protein; 4 g carbohydrate; 0 g fat; 150 mg sodium

Turn Off the Fat Genes

✳ Pasta Salad ✳

Makes about 8 cups

This fat-free pasta salad is delicious hot or cold. It is prepared with water-packed artichokes, which are available in most supermarkets.

4 ounces pasta spirals (2 cups uncooked pasta)
6 sundried tomatoes
½ cup finely chopped green onion
½ red bell pepper, diced
1 15-ounce can artichoke hearts, drained and quartered
¼ cup finely chopped fresh parsley
¼ cup chopped fresh basil
1 15-ounce can kidney beans, drained
1 cup fat-free Italian salad dressing

Cook the pasta in boiling water until it is just tender. Rinse with cold water, drain, and place in a large bowl. Soften the tomatoes by soaking them in ½ cup boiling water for 10 to 15 minutes. Drain and chop. Add to the pasta along with the remaining ingredients. Toss to mix.

Per cup: 92 calories; 3.5 g protein; 19 g carbohydrate; 0 g fat; 155 mg sodium

✳ South of the Border Salad ✳

Makes about 5 cups

This salad offers a cool and crunchy contrast to Mexican food or to any other spicy cuisine. It may be made in advance and keeps well.

1 large carrot, peeled and cut into ¼-inch rounds
1 medium jicama, peeled and diced (about 2 cups)
1 red bell pepper, seeded and diced
1 small sweet onion, thinly sliced (about ½ cup)
2 tablespoons finely chopped cilantro (optional)
3 tablespoons seasoned rice vinegar
3 tablespoons lemon juice
2 tablespoons Sesame Salt (page 207)

Combine the carrot, jicama, bell pepper, onion, and cilantro in a salad bowl. In a small bowl, combine the remaining ingredients. Pour over the vegetables and toss to mix.

Per ½ cup: 46 calories; 1 g protein; 9 g carbohydrate; 1 g fat; 121 mg sodium; 0 mg cholesterol

Hoppin' John Salad

Makes about 5 cups

2 cups cooked black-eyed peas (1 cup dry), or 1 15-ounce can, drained
1½ cups cooked brown rice (½ cup uncooked)
½ cup finely sliced green onions
1 celery stalk, thinly sliced (about ½ cup)
1 tomato, diced
2 tablespoons finely chopped fresh parsley

Combine the above ingredients in a mixing bowl.

¼ cup lemon juice
1 tablespoon olive oil

¼ teaspoon salt
1–2 garlic cloves, crushed

Mix together the vinaigrette ingredients and pour over the salad. Toss gently.

Chill 1 to 2 hours if time permits.

Per ½ cup: 84 calories; 3 g protein; 14 g carbohydrate; 1.5 g fat; 168 mg sodium

❋ California Waldorf Salad ❋

Makes about 6 cups

3 crisp apples, scrubbed and diced
2 carrots, julienned or grated
½ cup raisins
¼ cup chopped walnuts
½ cup Vicki's Tofu Mayo (page 199) or Fat-free
   Nayonaise
1 tablespoon lemon juice

Combine all ingredients in a salad bowl and toss to mix.

Per ½ cup: 76 calories; 2 g protein; 13 g carbohydrate; 1.5 g fat; 79 mg sodium

✻ Simple Vinaigrette ✻

Makes ½ cup

Seasoned rice vinegar makes a delicious salad dressing all by itself or with mustard and garlic.

½ cup seasoned rice vinegar
1–2 teaspoons stone-ground or Dijon-style mustard
1 clove garlic, pressed

Whisk all ingredients together. Use as a dressing for salads and steamed vegetables.

Per tablespoon: 14 calories; 0 g protein; 3 g carbohydrate; 0 g fat; 310 mg sodium

✻ Piquant Dressing ✻

Makes ½ cup

This dressing will be as spicy as the salsa you use to make it.

¼ cup seasoned rice vinegar
¼ cup salsa
1 garlic clove, pressed

Whisk all ingredients together.

Per tablespoon: 12 calories; 0 g protein; 3 g carbohydrate; 0 g fat; 210 mg sodium

Balsamic Vinaigrette

Makes ¼ cup

Balsamic vinegar has a mellow flavor that is perfect for salads.

2 tablespoons balsamic vinegar
2 tablespoons seasoned rice vinegar
1 garlic clove, crushed

Whisk all ingredients together.

Per tablespoon: 6 calories; 0 g protein; 1.5 g carbohydrate; 0 g fat; 99 mg sodium

Raspberry or Blackberry Vinaigrette

Makes ¼ cup

This dressing adds a hint of fruitiness to salads.

2 tablespoons raspberry or blackberry vinegar
2 tablespoons seasoned rice vinegar

Whisk all ingredients together.

Per tablespoon: 6 calories; 0 g protein; 1.5 g carbohydrate; 0 g fat; 99 mg sodium

✻ Curry Dressing ✻

Makes ½ cup

This spicy dressing is delicious on salad, steamed vegetables, or cooked grains.

3 tablespoons seasoned rice vinegar
3 tablespoons frozen apple juice concentrate
2 teaspoons stone-ground mustard
1 teaspoon soy sauce
½ teaspoon curry powder
¼ teaspoon black pepper

Whisk all ingredients together.

Per tablespoon: 9 calories; 0 g protein; 2 g carbohydrate; 0 g fat; 151 mg sodium

✻ Creamy Dill Dressing ✻

Makes about 1½ cups

This rich-tasting, creamy dressing has no added oil. Its creaminess comes from tofu.

8 ounces (1 cup) reduced-fat tofu
1 teaspoon garlic granules or powder
½ teaspoon dill weed
¼ teaspoon salt
2 tablespoons lemon juice
3 tablespoons seasoned rice vinegar
1 tablespoon cider vinegar

Combine all ingredients in a food processor or blender and blend until completely smooth. Store any extra dressing in an airtight container in the refrigerator.

Per tablespoon: 16 calories; 2 g protein; 2 g carbohydrate; 0.4 g fat; 92 mg sodium

SANDWICHES, WRAPS, AND ROLLS

❋ Garbanzo Salad Sandwich ❋

Makes 4 sandwiches

1 cup cooked garbanzo beans, drained
1 celery stalk, finely sliced
1 green onion, finely chopped, including green top
2 tablespoons Vicki's Tofu Mayo (page 199) or Fat-free Nayonaise
2 teaspoons stone-ground mustard
2 tablespoons sweet pickle relish
¼ teaspoon salt (omit if using canned beans)
8 slices whole wheat bread, or 4 pieces of pita bread
4 lettuce leaves
4 tomato slices

Mash the garbanzo beans with a fork or potato masher, leaving some chunks. Add the celery, green onion, Tofu Mayo, mustard, and relish. Add salt to taste. Spread on whole wheat bread or serve in pita bread with lettuce and sliced tomatoes.

Per sandwich: 182 calories; 7 g protein; 32 g carbohydrate; 3 g fat; 337 mg sodium

Tempeh Salad Sandwich

Makes 6 sandwiches

Tempeh is a soy food sold in natural food stores. Tempeh salad can also be served on a bed of lettuce and garnished with tomatoes.

- 8 ounces tempeh
- 3 tablespoons Vicki's Tofu Mayo (page 199) or Fat-free Nayonaise
- 2 teaspoons prepared mustard
- 2 green onions, chopped, including green tops
- 1 stalk celery, diced
- 1 tablespoon pickle relish
- 1/4 teaspoon salt
- 12 slices whole wheat bread
- 6 lettuce leaves
- 6 tomato slices

Steam the tempeh for 20 minutes. Remove from heat. When cool enough to handle, grate it and mix with the mayo, mustard, onions, celery, pickle relish, and salt. Cover and chill if time allows. Serve on whole wheat bread with lettuce and sliced tomatoes.

Per sandwich: 247 calories; 12 g protein; 35 g carbohydrate; 6 g fat; 407 mg sodium

Missing Egg Sandwich

Makes 4 sandwiches

- 1/2 pound firm reduced-fat tofu, mashed
- 1 green onion, finely chopped, including green top

2 tablespoons pickle relish
2 tablespoons Vicki's Tofu Mayo (page 199) or
   Fat-free Nayonaise
2 teaspoons stone-ground mustard
2 teaspoons soy sauce
¼ teaspoon each: cumin, turmeric, garlic powder
12 slices whole wheat bread
6 lettuce leaves
6 tomato slices

Combine all ingredients. Adjust seasonings if necessary. Serve on whole wheat bread with lettuce and sliced tomatoes.

Per sandwich: 177 calories; 11 g protein; 23 g carbohydrate; 4 g fat; 297 mg sodium

❋ VegiBurger Patties ❋

Makes sixteen 3-inch patties

These versatile patties may be topped with ketchup, barbecue sauce, or gravy or served in whole wheat buns with all the fixings.

¾ cup bulgur wheat
1 cup boiling water
1 small onion, finely chopped
1 medium carrot, shredded or finely chopped
2 stalks celery, finely chopped
1 pound mushrooms, finely chopped
½ cup walnuts, finely chopped
⅓ cup potato flour
½ teaspoon salt
½ teaspoon garlic powder

¼ teaspoon black pepper

1 tablespoon reduced-sodium soy sauce

Ketchup, barbecue sauce, or gravy for topping

Place the bulgur in a large bowl and pour the boiling water over it. Soak until the bulgur is tender and most of the water is absorbed, about 15 minutes.

Heat 2 tablespoons of water in a nonstick skillet and add the onion, carrot, and celery. Cook over medium-high heat for 3 minutes, stirring often. Stir in the mushrooms and continue cooking, stirring occasionally, until the vegetables are soft and the mushrooms are brown, about 5 minutes.

Drain any excess water off the bulgur. Add the vegetables along with the remaining ingredients and stir for 1 to 2 minutes, until the mixture holds together and can be formed into patties.

Form 3-inch patties, using about ¼ cup of the mixture for each. Mist a nonstick skillet with a vegetable oil spray and cook the patties until lightly browned, about 3 minutes per side.

NOTE: Patties may be frozen after they have been cooked. To reheat, place frozen patty in a toaster oven at 375°F for 5 minutes, or until heated through. To reheat in a microwave, wrap loosely in plastic and cook 2 to 3 minutes.

Per patty: 76 calories; 2 g protein; 12 g carbohydrate; 2 g fat; 118 mg sodium

✱ Vegetarian Reuben Sandwich ✱

Makes 6 sandwiches

Seitan ("say-tan"), also called wheat meat, is a high-protein, fat-free food with a meaty texture and flavor. Look for it in the deli case or freezer case of your natural food store.

1 onion, chopped
2 garlic cloves, minced
1½ cups sauerkraut
1 teaspoon paprika
1 teaspoon caraway seeds
½ teaspoon thyme
¼ teaspoon black pepper
1 8-ounce package seitan, drained and thinly
    sliced
12 slices rye bread
Stone-ground mustard
Vicki's Tofu Mayo (page 199) or Fat-free
    Nayonaise
2 tomatoes, sliced

Heat ½ cup of water in a large nonstick skillet and cook the onion and garlic, stirring occasionally, until the onion begins to brown, about 8 minutes. Stir in the sauerkraut, paprika, caraway seeds, thyme, and black pepper. Cook 5 minutes, stirring occasionally. Add the seitan slices and cook, stirring often, until thoroughly hot. Toast the bread.

To assemble the sandwiches, spread the bread with mustard and Tofu Mayo or Fat-free Nayonaise. Top with the sauerkraut mixture and tomato slices.

Per sandwich: 216 calories; 13 g protein; 32 g carbohydrate; 2 g fat; 570 mg sodium

✳ Tofu, Lettuce, and Tomato ✳ Sandwich (TLT)

Makes 2 sandwiches

Baked tofu comes in a variety of flavors and is sold in natural food stores and some supermarkets. Look for it in the deli section.

> 1 package baked tofu
> 4 slices whole wheat or rye bread
> 1–2 tablespoons stone-ground mustard
> 1–2 tablespoons Vicki's Tofu Mayo (page 199) or
>   Fat-free Nayonaise
> 6 lettuce leaves
> 6 tomato slices

Cut the tofu into ⅛-inch-thick slices. Toast the bread and spread it lightly with mustard and Tofu Mayo or Fat-free Nayonaise. Top with a slice of tofu, lettuce, tomato, and bread.

Per sandwich: 194 calories; 13 g protein; 22 g carbohydrate; 6 g fat; 320 mg sodium; 0 mg cholesterol

✳ Thai Wraps ✳

Makes 8 wraps

Thai Wraps are a delicious way to get your vegetables when you're on the run.

> 1 tablespoon peanut butter
> 2 tablespoons soy sauce
> 1 small onion, chopped
> 1 carrot, thinly sliced

1 celery stalk, thinly sliced

2 cups mushrooms, sliced

½ pound firm tofu, cut into ½-inch cubes

1½ teaspoons curry powder

½ red bell pepper, diced

½ cup cilantro, chopped (optional)

2 cups kale, finely chopped

6 large flour tortillas

2 cups cooked brown rice (page 270)

6 tablespoons Apple Chutney (page 207) or
Plum Sauce (page 205)

Mix the peanut butter with 3 tablespoons of water. Set aside.

Heat ½ cup water and the soy sauce in a large nonstick skillet. Add the onion, carrot, celery, and mushrooms and cook 5 minutes, stirring occasionally. Stir in the tofu and cook over medium-high heat, stirring often, until the vegetables are tender, about 5 minutes. Stir in the curry powder, red bell pepper, cilantro (if using), kale, and peanut butter mixture. Cover and cook until the kale is tender, about 5 minutes.

Heat the tortillas in a dry skillet until soft. Place about ½ cup of the vegetable mixture along the center of the tortilla. Top with ¼ cup of brown rice and 2 teaspoons of Plum Sauce or Apple Chutney. Roll the tortilla around the filling.

Per wrap: 246 calories; 10 g protein; 44 g carbohydrate; 3 g fat; 243 mg sodium

The Recipes

❋ Garbanzo Wraps ❋

Makes 6 to 8 wraps

These roll-ups are a portable meal in an edible wrapper, perfect
for lunches or picnics.

1 carrot, scrubbed
1–2 garlic cloves
1 15-ounce can garbanzo beans, drained
1 tablespoon tahini (sesame seed butter)
3 tablespoons lemon juice
¼ teaspoon cumin
¼ teaspoon paprika
6 flour tortillas or chapatis
2 cups cooked brown rice (page 270)
3 cups (approximately) prewashed salad mix
1 cup chopped cilantro (optional)

Finely chop the carrot and garlic in a food processor or by
hand. Add the drained beans, tahini, lemon juice, cumin, and
paprika and process until smooth, or mash until well mixed.

Spread about ½ cup of the garbanzo mixture along the center of a tortilla and top it with about ⅓ cup of the brown rice
and salad mix. Sprinkle with chopped cilantro if desired. Roll
the tortilla firmly around the filling.

Per wrap: 249 calories; 9 g protein; 46 g carbohydrate; 3 g fat; 252 mg sodium

Turn Off the Fat Genes

❋ Portabello and Red Pepper Wraps ❋

Makes 6 wraps

4 large firm portabello mushrooms
2 teaspoons toasted sesame oil
2 tablespoons red wine
1 tablespoon soy sauce
1 tablespoon balsamic vinegar
2 garlic cloves, minced
6 flour tortillas
1 cup roasted red peppers, cut into strips
4 cups finely shredded Napa cabbage
⅓ cup Plum Sauce (page 205)

Clean the mushrooms and trim off the stems. Combine the sesame oil, wine, soy sauce, vinegar, and minced garlic in a large nonstick skillet. Heat until the mixture bubbles; then place the mushrooms into the pan with the stem side up. Reduce the heat to medium, cover, and cook 3 minutes. If the pan becomes dry, add 2 to 3 tablespoons of water. Turn the mushrooms, cover, and cook until tender when pierced with a sharp knife, about 5 minutes. Cut the mushrooms into ½-inch-wide strips. Heat the tortillas in a dry skillet or microwave. To assemble, lay several pieces of mushroom mixture on each tortilla. Top with strips of red pepper, shredded cabbage, and Plum Sauce. Wrap the tortilla around the filling and serve.

Per wrap: 185 calories; 6 g protein; 32 g carbohydrate; 3 g fat; 177 mg sodium

❋ Burritos Supremos ❋

Makes 4 burritos

Burritos are one of the easiest wraps and can be eaten hot or cold.

> 1 15-ounce can vegetarian refried beans
> 4 flour tortillas
> 1 cup cooked brown rice (page 270)
> 1 cup shredded romaine lettuce
> 1 tomato, diced
> 2 green onions, sliced
> ½ cup salsa

Heat the beans in a saucepan or microwave. Heat the tortillas in a dry skillet until they are warm and soft. You can also wrap them loosely in a plastic bag and heat them in the microwave for 10 to 15 seconds.

Spread about ⅓ cup of the beans along the center of each tortilla, then top with rice, lettuce, tomato, onions, and salsa. Roll the tortilla around the filling. Repeat with remaining tortillas.

Per burrito: 327 calories; 14 g protein; 62 g carbohydrate; 2 g fat; 210 mg sodium

❋ Quickie Quesadillas ❋

Makes about 8 quesadillas

> 1 15-ounce can garbanzo beans
> ½ cup water-packed roasted red pepper
> 3 tablespoons lemon juice
> 1 tablespoon tahini (sesame seed butter)
> 1 garlic clove, peeled

Turn Off the Fat Genes

¼ teaspoon cumin
8 corn tortillas
½ cup chopped green onions
2 tomatoes, diced
½–1 cup salsa

Drain the garbanzo beans and place them in a food processor or blender with the roasted peppers, lemon juice, tahini, garlic, and cumin. Process until very smooth, 1 to 2 minutes. Spread a tortilla with a thin layer of the garbanzo mixture (2 to 3 tablespoons) and place it in a large nonstick skillet over medium heat. Sprinkle with some of the chopped green onions, diced tomatoes, and salsa. Top it with a second tortilla and cook until the bottom tortilla is warm and soft, 2 to 3 minutes. Turn and cook the second side for another minute. Remove from pan and cut in half. Repeat with remaining tortillas.

Per quesadilla: 137 calories; 5 g protein; 23 g carbohydrate; 2.4 g fat; 115 mg sodium

Pita Pizzas

Makes 6 pizzas

Whole wheat pita bread makes a perfect crust for individual pizzas. Serve Pita Pizzas for quick meals or snacks. You can make them almost instantly if you keep some pizza sauce and chopped vegetables in the refrigerator.

1 15-ounce can tomato sauce
1 6-ounce can tomato paste
1 teaspoon garlic granules or powder
½ teaspoon each: basil, oregano, thyme
2 green onions, thinly sliced

1 red bell pepper, diced
1 cup chopped mushrooms
6 pieces pita bread

Preheat the oven to 375° F. Combine the tomato sauce, tomato paste, garlic granules, and herbs. Prepare the vegetables as directed. For each pizza, turn a piece of pita bread upside down and spread it with sauce. Top with chopped vegetables. Place on a cookie sheet and bake until the edges are lightly browned, about 10 minutes.

NOTE: You will only need about half the sauce for 6 pizzas. Refrigerate or freeze the remainder for use at another time.

Per pizza: 185 calories; 7 g protein; 35 g carbohydrate; 2 g fat; 337 mg sodium

California Nori Rolls

Makes 3 rolls

A nori roll is a vegetarian sushi that is served unsliced, like a burrito. Of course, you can also cut it into ½-inch-thick slices for appetizers. Serve with extra pickled ginger and wasabe (horseradish paste) if desired. Baked tofu is sold in natural food stores and some supermarkets.

1 cup short grain brown rice
¼ teaspoon salt
3 cups water
¼ cup seasoned rice vinegar
1 cup grated carrot
1 cup grated cucumber
1 cup grated baked tofu
1 green onion, chopped or cut into thin strips

¼ avocado, cut into 8 slices (optional)
⅓ cup (approximately) pickled ginger
3 sheets nori

Pickled ginger for serving
Wasabe for serving, if desired (very hot!)

Place the brown rice in a saucepan with the salt and water. Cover and bring to a simmer, then cook until the rice is soft and all the water has been absorbed, about 1 hour. Remove from heat. When cool, stir in the seasoned rice vinegar. Set aside. Grate the carrot, cucumber, and baked tofu. Chop or slice the green onion.

To assemble the rolls, place a sheet of nori on a bamboo sushi mat. Spread about ¾ cup of the cooled rice in a thin, even layer on the sheet of nori, leaving a 1-inch band along the top of the sheet uncovered. Arrange a small amount of each of the fillings, including the avocado (optional) and pickled ginger, over the rice. To form the roll, hold the filling ingredients in place with your fingertips, and use your thumbs to lift the bottom edge of the mat, so that the edge of the nori nearest you is lifted over to meet the top edge. Moisten the top edge and use it as a "flap" to seal the roll. Use your hands to gently shape the roll, then let it sit on its seam to seal.

Per ½ roll (with avocado): 170 calories; 7 g protein; 25 g carbohydrate; 3 g fat; 227 mg sodium
Per ½ roll (without avocado): 149 calories; 6 g protein; 24 g carbohydrate; 2 g fat; 225 mg sodium

⁕ Portuguese Kale Soup ⁕

Makes about 12 cups

This hearty soup is a delicious way to enjoy calcium-rich kale. The recipe calls for Yves Veggie Breakfast Links, which are sold in natural food stores and many supermarkets.

1 tablespoon olive oil
1 onion, chopped
1 large carrot, scrubbed and diced
2 celery stalks, thinly sliced
2 russet potatoes, scrubbed and diced
3 garlic cloves, minced
2 teaspoons dried oregano
1 teaspoon dried basil
¼ teaspoon black pepper
¼ teaspoon salt
1 15-ounce can cannelini beans
1 quart Imagine Vegetable Broth or other vegetable stock
1 7-ounce package Yves Veggie Breakfast Links
½ pound chopped kale

Heat the oil in a large pot and add the onion, carrot, and celery. Cook over high heat, stirring often, until the onion is soft, about 3 minutes. Reduce the heat to medium and stir in the potatoes, garlic, seasonings, and ¼ cup of water. Cover and cook, stirring often, for 5 minutes. Stir in the cannelini beans and vegetable broth. Cover and simmer until the potatoes are just tender, about 10 minutes. Cut the breakfast links into bite-sized chunks and add them, along with the kale. Cover and simmer until the kale is tender, about 5 minutes.

Per cup: 134 calories; 7 g protein; 22 g carbohydrate; 1 g fat; 344 mg sodium

Spicy Thai Soup

Makes about 6 cups

Serve this simple vegetable soup with Ginger Noodles (page 271).

- 1 quart vegetable broth*
- 1 tablespoon finely chopped fresh ginger
- 2 teaspoons minced garlic
- ½–1 jalapeño pepper, seeded and finely chopped (or more to taste)
- 1 cup sliced mushrooms
- 1 cup broccoli, cut into bite-sized florets
- 1 cup (packed) bok choy, finely chopped
- 1 green onion, finely chopped, including top
- 1 tablespoon finely chopped cilantro

Combine the vegetable broth, ginger, garlic, and jalapeño pepper in a pot and bring to a boil. Add the mushrooms and simmer for 2 minutes. Add the broccoli and bok choy. Simmer until the broccoli is tender but still bright green and crisp, 3 to 4 minutes. Do not overcook! Stir in the green onion and chopped cilantro. Serve immediately.

Per cup: 23 calories; 1 g protein; 4 g carbohydrate; 0.5 g fat; 340 mg sodium

*Vegetable broth is available in liquid and powdered forms. Look for it in natural food stores or your supermarket.

✖ Yam and Corn Chowder ✖

Makes about 8 cups

1 large or 2 small yams
2 teaspoons extra virgin olive oil
1 onion, chopped
2 garlic cloves, minced
1 teaspoon curry powder
½ teaspoon ginger
¼ teaspoon cinnamon
¼ teaspoon coriander
½ teaspoon salt
1 tablespoon maple syrup or other sweetener
2 cups water or vegetable stock
2 cups Pacific Cream Flavored Sauce Base or
    soy milk
1 15-ounce can corn, including liquid
1 tablespoon lemon juice

Scrub the yam, peel if desired, and cut into ½-inch cubes. Set aside.

Heat the oil in a large pot, then add the onion and garlic and cook over medium heat until the onion is soft, about 3 minutes. Add the curry powder, ginger, cinnamon, coriander, and salt. Cook over medium heat for 2 minutes, stirring constantly. Stir in the maple syrup, water or vegetable stock, and the yams. Simmer until the yams are tender, about 15 minutes.

Transfer to a blender and puree in small batches, adding some of the soup base or soy milk to each batch. Return to the pot and add the corn. Heat over a medium flame until hot and steamy (do not let it boil), about 10 minutes. Stir in the lemon juice before serving.

Per cup: 137 calories; 3.5 g protein; 26 g carbohydrate; 2.5 g fat; 169 mg sodium

❈ Double Potato Soup ❈

Serves 8

Sweet potatoes and russets combine to make a deliciously different soup.

> 1 large onion, chopped
> 2 garlic cloves, minced
> 2 russet potatoes, peeled and diced
>   (about 4 cups)
> 1 large or 2 small sweet potatoes, peeled and
>   diced (about 4 cups)
> 4 cups vegetable broth
> 2 cups unsweetened soy milk
> 1 bunch kale (about 4 cups chopped)
> ½ teaspoon curry powder

Heat ½ cup of water in a large pot and cook the onion and garlic until the onion is soft, about 5 minutes. Add the potatoes and vegetable broth. Cover and simmer until the potatoes are tender when pierced with a fork, about 15 minutes. Remove about 3 cups of the potato mixture and puree with the soy milk in a blender until smooth. Set aside. Add the kale and curry powder to the pot, cover and cook over medium heat until kale is tender, about 5 minutes. Stir in the potato puree and continue to heat until steamy.

Per cup: 127 calories; 3 g protein; 27 g carbohydrate; 1 g fat; 32 mg sodium

❊ Quick & Creamy Beet Soup ❊

Makes about 3 cups

> 1 15-ounce can diced beets
> 1 cup soy milk
> 2 tablespoons apple juice concentrate
> 1 teaspoon balsamic vinegar
> ½ teaspoon dried dill weed

Place the diced beets with their liquid into a blender and add the remaining ingredients. Blend on high speed until completely smooth, 2 to 3 minutes. Transfer to a medium saucepan and heat gently until steamy.

Per cup: 67 calories; 2 g protein; 14 g carbohydrate; 0.5 g fat; 80 mg sodium

❊ Autumn Stew ❊

Makes about 10 cups

This colorful stew is a culinary celebration of the abundance of autumn. For special occasions, serve it in a pumpkin that has been hollowed out and baked until just tender.

> 1 tablespoon soy sauce
> 1 onion, chopped
> 1 red bell pepper, diced
> 4 large garlic cloves, minced
> 1 pound (about 4 cups diced) butternut squash
> 1 15-ounce can crushed tomatoes
> 1½ teaspoons oregano
> 1 teaspoon chili powder

½ teaspoon cumin

¼ teaspoon black pepper

1 15-ounce can kidney beans, undrained

1 15-ounce can corn, undrained (or 2 cups frozen)

Heat ½ cup of water and the soy sauce in a large pot. Add the onion, bell pepper, and garlic. Cook over medium heat until the onion is soft and most of the water has evaporated, about 5 minutes. Peel the squash, then cut it in half. Scoop out the seeds and discard. Cut the squash into ½-inch cubes and add it to the cooked onions along with the crushed tomatoes, 1 cup of water, the oregano, chili powder, cumin, and pepper. Cover and simmer until the squash is just tender when pierced with a fork, about 20 minutes. Add the kidney beans and corn with their liquid and cook 5 minutes longer.

Per cup: 105 calories; 4 g protein; 21 g carbohydrate; 0 g fat; 214 mg sodium

❈ Mexican Corn Chowder ❈

Serves 8

2–3 cups peeled and diced russet potatoes

2 cups water or vegetable stock

1 yellow onion, chopped

2 garlic cloves, minced

1 red bell pepper, diced

1 teaspoon ground cumin

1 teaspoon basil

½ teaspoon salt

¼ teaspoon turmeric

¼ teaspoon black pepper
2 15-ounce cans corn
1 4-ounce can diced green chilies
1–2 cups soy milk

Peel and dice the potatoes and put them in a pot with 2 cups of water or vegetable stock. Cover and cook until tender, about 20 minutes. In a separate pan, heat ½ cup of water and cook the onion, garlic, and bell pepper for 5 minutes. Add the cumin, basil, salt, turmeric, and black pepper. Continue to cook until the onion is very soft, about 5 minutes or longer. When the potatoes are tender, mash them in their water and combine with the onion mixture. Blend one of the cans of corn, with its liquid, until smooth, 2 to 3 minutes, then add it along with the remaining can of corn, the diced chilies, and 1 cup of the soy milk to the potato mixture. Stir to mix. Add more soy milk if a thinner soup is desired. Heat gently until hot and steamy.

Per cup: 112 calories; 3 g protein; 22 g carbohydrate; 1 g fat; 118 mg sodium

❈ Gazpacho ❈

Makes about 12 cups

This Spanish soup is perfect for a hot summer day.

2 cucumbers, peeled, seeded, and diced
1 green bell pepper, seeded and diced
3 ripe tomatoes, diced
½ cup finely chopped red onion
3 garlic cloves, crushed
¾ cup salsa
¾ cup roasted red peppers, finely chopped

Turn Off the Fat Genes

8 cups vegetable juice (like V-8) or tomato juice
1 teaspoon finely minced jalapeño pepper (more
    or less to taste)

Combine all ingredients and chill before serving.

Per cup: 53 calories; 1 g protein; 12 g carbohydrate; 0 g fat; 382 mg sodium

❋ Root Soup ❋

Serves 6 to 8

Beets and carrots combine to make a glorious crimson soup
that is simply delicious.

3 medium beets, chopped
3 medium carrots
1½ cups rice milk or soy milk
2–3 tablespoons apple juice concentrate
2 teaspoons balsamic vinegar
1 teaspoon dried dill weed

Cut the tops and roots off the beets, then scrub them and peel
off any rough portions. Cut into 1-inch chunks (you should
have about 3 cups) and place them in a soup pot. Cut the tops
off the carrots, scrub them, and cut them into 1-inch pieces
(you should have about 3 cups). Add the carrots to the pot,
along with 1½ cups of water, then cover and simmer until the
beets and carrots are tender when pierced with a knife, about
15 minutes.

Transfer about 2 cups of the beets and carrots to a blender,
along with some of the cooking liquid and some of the rice or
soy milk. Blend until completely smooth. Be sure to hold the

The Recipes

lid on tightly and start the blender on a low speed. Repeat with the remaining beets and carrots, adding some of the cooking liquid and some of the rice or soy milk to each batch. Add the apple juice concentrate, vinegar, and dill weed to the final batch. Return the blended soup to the pot and heat gently, until hot and steamy.

VARIATION: This soup is also delicious cold. Chill completely before serving.

Per cup: 64 calories; 2 g protein; 13 g carbohydrate; 0.5 g fat; 273 mg sodium

✱ Split Pea Soup ✱

Makes about 12 cups

This soup practically makes itself. Simply combine all of the ingredients in a large pot and let it simmer for 1 hour. If you prefer to use a crockpot, cook on high for 6 to 8 hours.

2 cups split peas, rinsed
8 cups water
1 onion, chopped
2 garlic cloves, minced
1 large carrot, diced
2 celery stalks, sliced
2 russet potatoes, diced
1 bay leaf (optional)
1 teaspoon marjoram
1 teaspoon basil
¼ teaspoon dried red pepper flakes or
    pinch cayenne (optional)
¼ teaspoon black pepper
1 teaspoon salt

Turn Off the Fat Genes

Combine all of the ingredients in a large pot. Bring to a simmer, then cover loosely and cook until the peas are tender, about 1 hour. Remove the bay leaf. For a thicker, more uniform soup, use a potato masher to mash some of the peas and vegetables.

Per cup: 126 calories; 7 g protein; 24 g carbohydrate; 0 g fat; 190 mg sodium

✺ Summer Vegetable Stew ✺

Makes about 10 cups

2 teaspoons olive oil
2 onions, chopped
3 Japanese eggplant, cut into ¼-inch-thick slices
1 green bell pepper, diced
5 large garlic cloves, minced
1 12-ounce jar water-packed roasted red peppers, including liquid
3 small zucchini, sliced
2 cups chopped fresh basil
1 15-ounce can navy beans or cannelini beans
½ teaspoon salt
¼ teaspoon black pepper

Heat the oil in a large skillet or pot. Add the onions and ¼ cup of water. Cook over medium-high heat, stirring often, until the onions are lightly browned, about 5 minutes. Add the eggplant, bell pepper, and garlic, then cover and cook for 5 minutes, stirring occasionally. Coarsely chop the roasted red peppers, then add them with their liquid. Add the zucchini and basil, then cover and cook over medium heat, stirring occasionally, for 3 minutes. Stir in the navy or cannelini beans with their liquid, the salt, and pepper. Cover and cook until the zucchini is just tender, about 3 minutes.

Per cup: 97 calories; 3.5 g protein; 18 g carbohydrate; 1 g fat; 171 mg sodium

VEGETABLES

Beets in Dill Sauce

Makes about 4 cups

These beets are delicious hot or chilled.

- 4 medium-size beets
- 2 tablespoons lemon juice
- 1 tablespoon stone-ground mustard
- 1 tablespoon cider vinegar
- 1 tablespoon apple juice concentrate
- 1 teaspoon dried dill weed, or 1 tablespoon fresh dill, chopped

Wash and peel the beets, then slice them into ¼-inch-thick rounds. Steam over boiling water until tender when pierced with a fork, about 20 minutes. Mix the remaining ingredients in a serving bowl. Add the beets and toss to mix. Serve immediately, or chill before serving.

Per ½ cup: 33 calories; 1 g protein; 7 g carbohydrate; 0 g fat; 65 mg sodium

Sesame Kale

Makes about 4 cups

1 bunch kale (about ½ pound)
2 tablespoons Sesame Salt (page 207)

Rinse the kale and remove the stems. Cut or tear the leaves into bite-sized pieces. Steam over boiling water until tender, about 5 minutes. Transfer to a serving bowl and toss with Sesame Salt.

Per ½ cup: 31 calories; 1 g protein; 4 g carbohydrate; 1 g fat; 49 mg sodium

Summer Succotash

Makes about 8 cups

Serve in place of broccoli with Red Beans and Rice (page 266).

2 teaspoons olive oil
1 large onion, chopped
2 garlic cloves, minced
1½ teaspoons dried basil
¼ teaspoon black pepper
1 large yellow zucchini, cut into ½-inch cubes (about 2 cups)
2 cups (packed) finely chopped fresh kale
2 cups frozen Italian green beans
1 15-ounce can corn, including liquid
2 tablespoons soy sauce

Heat the olive oil in a large skillet and sauté the onion and garlic until the onion is soft, about 5 minutes. Rub the basil between your hands to crush it, and add it to the onions along with the black pepper, zucchini, and chopped kale. Stir to mix, then cook 2 minutes. Add the green beans and the corn with its liquid. Cover and cook over medium heat, stirring occasionally, until the kale and zucchini are tender, about 5 minutes. Stir in soy sauce before serving.

Per ½ cup: 49 calories; 1.5 g protein; 9 g carbohydrate; 1 g fat; 85 mg sodium

✖ Mashed Potatoes with Black Beans ✖

Makes 6 to 8 cups

4 russet potatoes
10 green onions, chopped, including green tops
10 garlic cloves, minced
1 large carrot, grated
½ teaspoon salt
¼ teaspoon black pepper
1 15-ounce can black beans, drained
Fresh cilantro for garnish (optional)

Scrub the potatoes, peel if desired, and cut into quarters. Place in a pot with 1 cup of water. Cover and simmer until the potatoes are very soft, about 20 minutes. Heat ½ cup of water in a large nonstick skillet and add the green onions, garlic, and carrot. Cook over high heat, stirring frequently, until all the liquid has evaporated, about 5 minutes. Add the potatoes, with their cooking liquid, and mash, leaving some chunks. Be careful not to scratch the pan. Stir in the salt, pepper, and drained beans. Heat gently, stirring occasionally. Garnish with chopped cilantro if desired.

Per ½ cup: 83 calories; 2 g protein; 18 g carbohydrate; 0 g fat; 106 mg sodium

Turn Off the Fat Genes

Skillet Scalloped Potatoes

Makes about 8 cups

4 medium red potatoes
2 cups Pacific Cream Flavored Sauce Base
⅓ cup Sesame Seasoning (page 208)
1 bunch broccoli
Paprika

Scrub the potatoes, then steam them over boiling water until tender, about 30 minutes. Set aside to cool. When the potatoes are cool enough to handle, pour the sauce base into a large nonstick skillet. Slice the potatoes ¼ inch thick and arrange them evenly in the skillet. Sprinkle with 3 tablespoons of Sesame Seasoning. Bring to a simmer over medium-low heat and cook 5 minutes.

While the potatoes simmer, cut the broccoli into bite-sized florets and steam over boiling water until bright green and tender-crisp, 3 to 5 minutes. Spread evenly over the potatoes. Sprinkle with remaining Sesame Seasoning. Sprinkle with paprika. Continue to simmer, uncovered, until most of the liquid has evaporated, about 5 minutes. Serve hot.

Per ½ cup: 93 calories; 2 g protein; 16 g carbohydrate; 2 g fat; 122 mg sodium

⁂ Mashed Potatoes and Gravy ⁂

Makes 5 to 6 cups

Now you can enjoy this traditional favorite to your heart's content!

- 4 russet potatoes, peeled and diced
- 1 cup soy milk
- ¼ teaspoon onion powder
- ¼ teaspoon garlic powder
- ½ teaspoon salt
- 1½ tablespoons rice flour
- 2 tablespoons potato flour
- ½ cup Corn Butter (page 200)

Put the potatoes and 2 cups of water in a saucepan. Simmer until tender when pierced with a fork, about 10 minutes. Drain, reserving the liquid to make the gravy. Pour the soy milk into a blender. With the blender running, add the remaining ingredients and blend until completely smooth. Mash the potatoes, then add the sauce and mix well.

To make the gravy, pour the reserved potato cooking water into a measuring cup and add enough water to make 2 cups. Transfer to a blender and add the following ingredients:

- 1 tablespoon cashews
- 1 tablespoon onion powder
- ½ teaspoon garlic granules or powder
- 2 tablespoons cornstarch
- 2–3 tablespoons soy sauce

Blend until completely smooth, then transfer to a saucepan and cook over medium heat, stirring constantly, until thickened.

Mashed potatoes: per ½ cup: 76 calories; 2 g protein; 16 g carbohydrate; 1 g fat; 157 mg sodium
Gravy: per ¼ cup: 15 calories; 0.3 g protein; 2 g carbohydrate; 0.5 g fat; 63 mg sodium

✳ Potatoes and Kale ✳

Makes about 4 cups

Using Yukon Gold or Yellow Finn potatoes gives this dish a delicious buttery flavor without adding any fat.

> 8 small Yukon Gold or Yellow Finn potatoes
> 4 cups chopped kale
> 1 onion, thinly sliced
> 2 garlic cloves, minced
> ½ teaspoon each: black pepper, paprika
> 2 tablespoons reduced-sodium soy sauce

Scrub the potatoes and cut into cubes or wedges (3 to 4 cups). Steam until just tender when pierced with a fork, about 10 minutes. Rinse with cold water, drain, and set aside. Rinse the kale and remove the stems. Cut or tear the leaves into small pieces. Heat ½ cup of water in a large nonstick skillet and cook the onion and garlic over high heat, stirring occasionally, for 5 minutes. Reduce heat to medium and add the cooked potatoes, pepper, paprika, and soy sauce. Cook, turning frequently with a spatula, until the potatoes begin to brown, about 5 minutes. Spread the chopped kale over the potatoes and sprinkle with 2 tablespoons of water. Cover and cook, turning occasionally, until the kale is tender, about 10 minutes.

Per ½ cup: 74 calories; 4 g protein; 20 g carbohydrate; 0 g fat; 242 mg sodium

❈ Ratatouille ❈

Serves 6 to 8

This French vegetable dish celebrates the bounty of the summer garden. Serve it with crusty French bread and a crisp green salad.

2 onions, chopped
3 garlic cloves, minced
1 large eggplant, diced
1 15-ounce can crushed tomatoes
½ teaspoon each: basil, oregano, thyme, salt
¼ teaspoon black pepper
1 green bell pepper, diced
2 medium zucchini, sliced

Heat ½ cup of water in a large pot and add the onions and garlic. Cook over medium heat until the onions are soft, about 5 minutes. Stir in the eggplant, crushed tomatoes, and seasonings. Cover and cook, stirring frequently, until the eggplant is just tender, about 15 minutes. Stir in the bell pepper and zucchini. Cover and cook until tender, about 5 minutes.

Per ½ cup: 41 calories; 1 g protein; 9 g carbohydrate; 0 g fat; 62 mg sodium

❈ Curried Cauliflower with Peas ❈

Makes 5 to 6 cups

1 large cauliflower
1 teaspoon coriander
1 teaspoon whole mustard seed
½ teaspoon each: turmeric, cumin

Turn Off the Fat Genes

¼ teaspoon each: cayenne, cinnamon, ginger
⅛ teaspoon cardamom
½ cup vegetable stock or water
1 large onion, chopped
2 cups frozen peas
¼ teaspoon salt

Rinse the cauliflower, then divide it into bite-sized florets. Set aside. Toast the spices in a small dry skillet until they are fragrant and just begin to darken, about 1 minute. Heat the vegetable stock or water in a large skillet. Add the onion and cook until it is soft, about 5 minutes. Stir in the cauliflower and spices, then cover and cook over medium heat until the cauliflower is tender when pierced with a fork, about 5 minutes. Stir in the peas and cook until hot, another minute or two. Add salt to taste.

Per ½ cup: 43 calories; 2 g protein; 8 g carbohydrate; 0 g fat; 92 mg sodium

❇ Carrots in Orange Sauce ❇

Makes about 3 cups

2 large carrots
¾ cup orange juice
1½ teaspoons cornstarch
1 tablespoon apple juice concentrate
1 teaspoon reduced-sodium soy sauce
¼ teaspoon dill weed

Scrub the carrots and cut into ¼-inch-thick slices. You should have 2 to 3 cups. Steam until just tender. Combine the remaining ingredients in a saucepan, whisk smooth, then bring to a

The Recipes

simmer over medium heat, stirring constantly. Cook until sauce is clear and slightly thickened. Add carrots and heat through. Serve hot or cold.

Per ½ cup: 32 calories; 0.4 g protein; 8 g carbohydrate; 0 g fat; 43 mg sodium

❊ Broccoli with Sesame Salt ❊

Makes about 4 cups

Here is a simple and delicious way to prepare broccoli. Serve it with cooked rice, couscous, pasta, or grain of your choice.

> 1 bunch broccoli
> 1 tablespoon Sesame Salt (page 207)

Rinse the broccoli, then remove the stem. Cut or break the florets into bite-sized pieces. Peel the stems and cut them into ¼-inch-thick rounds. Place on a steamer rack over boiling water, then cover and cook until just tender, about 5 minutes. The broccoli should still be bright green and slightly crisp. Transfer to a mixing bowl and sprinkle with Sesame Salt. Toss to mix.

Per ½ cup: 25 calories; 1 g protein; 3 g carbohydrate; 1 g fat; 46 mg sodium

❊ Oven Roasted Vegetables ❊

Serves 6 to 8

Roasting vegetables in a hot oven brings out their best flavor. For variety, try tossing them with different herbs.

Turn Off the Fat Genes

1 large onion, peeled
3 summer squash, about 3 cups (zucchini, crookneck, or scallop squash)
1 large red bell pepper, seeded
2 cups small, firm mushrooms
2 ears corn, cut into ½-inch-thick slices
2 teaspoons olive oil
2 teaspoons garlic granules
1 teaspoon thyme
½ teaspoon rosemary
½ teaspoon salt
½ teaspoon black pepper

Preheat oven to 500° F. Line one or two large baking dishes with baking parchment or aluminum foil. Cut the onion, squash, and bell pepper into large chunks and place in a large bowl. Clean the mushrooms and add to the bowl along with the corn. Sprinkle with olive oil and seasonings and toss to mix. Spread in a single layer in the baking dish or dishes. Bake in preheated oven until tender, 10 to 15 minutes.

Per ½ cup: 28 calories; 1 g protein; 5 g carbohydrate; 1 g fat; 70 mg sodium

Winter Squash

Makes 4 cups

Winter squash has so much going for it, including its naturally sweet, buttery flavor, its ease of preparation, and the fact that it is available most of the year. There are many different varieties, including butternut, acorn, kabocha, delicata, and sweet dumpling. Each has a slightly different flavor and texture. The fastest and easiest way to cook winter squash is to steam it. It can be eaten plain, stuffed (page 287), or tossed with the simple sauce that follows.

> 1 medium-size winter squash (butternut,
>   kabocha, delicata, etc.)
> ½ cup water
> 2 teaspoons reduced-sodium soy sauce
> 1 tablespoon maple syrup

Peel the winter squash and remove the seeds. Cut the squash into 1-inch cubes (about 4 cups) and place it into a large pot with the water, soy sauce, and maple syrup. Cover and simmer over medium heat until the squash is fork tender.

Per ½ cup: 46 calories; 1 g protein; 10 g carbohydrate; 0 g fat; 51 mg sodium

❈ Brussels Sprouts in Creamy Sauce ❈

Makes about 4 cups

The creamy sauce is made with Pacific Cream Flavored Sauce Base, a nondairy cream soup base made by Pacific Foods of Oregon and sold in natural food stores and many supermarkets. You can also use soy milk to make the sauce.

> 1 pound fresh brussels sprouts (about 4 cups)
> 1 small onion, sliced
> 2 tablespoons Sesame Salt (page 207)
> 1 tablespoon whole wheat pastry flour
> ¼ teaspoon salt
> ⅛ teaspoon celery seeds
> ⅛ teaspoon black pepper

1 cup Pacific Cream Flavored Sauce Base
or soy milk
⅛ teaspoon nutmeg

Trim the brussels sprouts, removing any wilted leaves. Put 1 inch of water in a large pot and bring to a boil. Add the brussels sprouts, cover, and cook 12 minutes. Drain and set aside.

In a large nonstick skillet, heat ½ cup of water. Add the onion and cook over high heat, stirring often, until the onion begins to brown and all the liquid has evaporated, about 5 minutes. Add ¼ cup of water, stir to loosen any stuck bits of onion, and continue cooking over high heat, stirring often until the onion is nicely browned, about 5 more minutes. Add a tablespoon or two of water whenever the onion begins to stick.

Lower heat to medium and stir in the Sesame Salt, flour, salt, celery seeds, and black pepper. Cook 1 minute, stirring constantly. Add the cream flavored base or soy milk. Bring to a simmer and cook 2 minutes, stirring constantly. Stir in the brussels sprouts and continue cooking 2 to 3 minutes. Sprinkle with nutmeg before serving.

Per ½ cup: 59 calories; 2 g protein; 9 g carbohydrate; 2 g fat; 154 mg sodium

❋ Steamed Vegetables with ❋ Sesame Salt

Makes 8 to 10 cups

There is a very good reason why steamed vegetables with brown rice has become legendary vegetarian fare: it is quick and simple to prepare and it tastes great. By varying the vegetables according to your taste and what's in season, you can easily create a regular dish that is ever-changing and always delicious. The guiding principle is to start steaming the longer-cooking vegetables first; then, when they are just barely tender, adding the quick-cooking vegetables like broccoli, cauliflower, peppers, and summer squash. Here is a recipe to get you started.

1 yellow onion, peeled
1 carrot, scrubbed
1 potato, scrubbed
1 yam or sweet potato, scrubbed
6 garlic cloves, peeled
1 cup button mushrooms, cleaned
2 cups broccoli or cauliflower florets
¼ cup Sesame Salt (page 207)

Brown rice for serving (page 270)

Cut the onion, carrot, potato, and yam into ½- to 1-inch chunks. Place in a pot on a vegetable steamer with the garlic and mushrooms. Cover and steam until the potatoes are just barely tender when pierced with a fork. Add the broccoli or cauliflower. Cover and cook until the broccoli is bright green and just tender, about 5 minutes. Transfer to a mixing bowl and sprinkle with Sesame Salt. Toss to mix. Serve on a bed of brown rice.

Per ½ cup of vegetables: 52 calories; 1 g protein; 10 g carbohydrate; 1 g fat; 34 mg sodium

❋ Green Beans with Garlic ❋

Makes about 2 cups

1 package frozen, or 1 pound fresh Italian green beans

1 teaspoon toasted sesame oil

8 garlic cloves, minced

2 tablespoons seasoned rice vinegar

1 tablespoon reduced-sodium soy sauce

¼ teaspoon black pepper

Cook the beans according to package directions or, if using fresh beans, steam until just tender, about 10 minutes. Set aside. Heat the oil in a nonstick skillet and sauté the garlic, stirring constantly, for 1 minute. Stir in the vinegar, soy sauce, 2 tablespoons of water, and the cooked beans. Sprinkle with pepper and cook, stirring constantly, until the mixture is very hot, about 2 minutes.

Per ½ cup: 61 calories; 2 g protein; 11 g carbohydrate; 1 g fat; 306 mg sodium

BEANS

❈ Quick Black Bean Chili ❈

Makes about 8 cups

This is a mild chili, delicious with brown rice and a green salad. It can also be used as a burrito filling if it is cooked until it gets thick. If you like a hotter chili, add some cayenne or fresh chopped jalapeños.

½ cup water

1 tablespoon reduced-sodium soy sauce

2 onions, chopped

4 garlic cloves, crushed

2 teaspoons oregano

½ teaspoon cumin

¼ teaspoon black pepper

1 4-ounce can diced chilies

1 15-ounce can crushed tomatoes

2 15-ounce cans black beans

The Recipes

¼ teaspoon salt
Fresh cilantro, chopped

Heat the water and soy sauce in a large pan and add the onions and garlic. Cook over medium heat until the onions are soft, about 5 minutes. Add the oregano, cumin, and black pepper and cook 3 minutes; then stir in the diced chilies, tomatoes, black beans, and salt. Simmer until thickened, 20 minutes or longer. Sprinkle with chopped cilantro.

Per ½ cup: 92 calories; 5 g protein; 18 g carbohydrate; 0 g fat; 174 mg sodium

❇ Red Lentil Curry ❇

Makes about 4 cups

This dish is easy to make and cooks quickly. If you can't locate red lentils, yellow split peas may be used.

3½ cups water or vegetable stock
1 onion, chopped
1 carrot, diced
1 celery stalk, sliced
1 cup red lentils
½ teaspoon salt
1 teaspoon mustard seeds
½ teaspoon turmeric
½ teaspoon cumin
½ teaspoon coriander
½ teaspoon ginger
⅛–¼ teaspoon cayenne

Heat ½ cup of the water or vegetable stock in a large pot. Add the onion, carrot, and celery. Cook over medium heat until the onion is soft and translucent, about 5 minutes. Sort through the red lentils to remove debris, rinse, and add to the pot along with the remaining water and the salt. Bring to a simmer. Cover loosely and cook until the lentils are tender, about 20 minutes (45 minutes for yellow split peas). Toast the spices in a dry skillet (be careful not to inhale the fumes) until the mustard seeds begin to pop. This will happen quickly. Add the toasted spices to the cooked lentils. Continue cooking over medium heat, stirring occasionally, until thickened, about 10 minutes.

Per ½ cup: 87 calories; 5 g protein; 16 g carbohydrate; 0.2 g fat; 143 mg sodium

✖ French Green Lentils ✖

Makes about 6 cups

French green lentils have a distinctive and delicious peppery flavor. Look for them in natural food stores and fine food markets. If brown lentils are used as a substitute, increase the cooking time to 50 minutes.

> 1 cup French green lentils
> 4 cups water or vegetable stock
> 1 large onion, chopped
> ½ cup chopped cilantro
> 1 teaspoon black mustard seeds
> 1 teaspoon turmeric
> 1 teaspoon cumin
> 1 teaspoon coriander
> ½ teaspoon ginger
> ⅛–¼ teaspoon cayenne
> ½ teaspoon salt

The Recipes

Rinse the lentils and place them in a large pot with the vegetable stock or water, the onion, and cilantro. Toast the spices, stirring them constantly, in a small dry skillet until they are fragrant and just begin to smoke, about 2 minutes. Add them to the lentils. Cover and simmer until the lentils are tender, about 30 minutes. Add salt to taste.

Per ½ cup: 55 calories; 4 g protein; 10 g carbohydrate; 0 g fat; 90 mg sodium

✖ Chili Beans ✖

Makes about 7 cups

These chili beans are delicious with Corn Bread (page 192), warm tortillas, or brown rice. A crisp green salad rounds out the meal beautifully.

1½ cups dry pinto beans
4 cups water
3 large garlic cloves, minced
½ teaspoon ground cumin
1 onion chopped
1 green bell pepper, diced
1 cup textured vegetable protein (optional)*
1 15-ounce can tomato sauce
1 cup corn, fresh or frozen
2 teaspoons chili powder
⅛ teaspoon cayenne (more for spicier beans)
½–1 teaspoon salt

---

*Textured vegetable protein (TVP) is a meatlike ingredient made from defatted soy flour. It is sold in natural food stores and many supermarkets (look for it in the bulk bins).

Sort through the beans to remove any debris, then rinse and soak for 6 to 8 hours in about 6 cups of cold water. Discard the water and rinse the beans. Place in a pot with 4 cups of fresh water, the garlic and cumin. Simmer until tender, about 1 hour. Heat ½ cup of water in a large skillet and add the onion, bell pepper, and textured vegetable protein (if using). Cook over medium heat, stirring often, until the onion is soft, about 5 minutes. Mix with the cooked beans, tomato sauce, corn, chili powder, and cayenne. Simmer at least 30 minutes. Add salt to taste.

Per ½ cup with TVP: 109 calories; 7 g protein; 19 g carbohydrate; 0 g fat; 95–171 mg sodium
Per ½ cup without TVP: 90 calories; 4 g protein; 17 g carbohydrate; 0 g fat; 93–169 mg sodium

❉ Baked Beans ❉

Makes about 8 cups

These beans may be "baked" on the stove top, in the oven, or in a crockpot. The longer they cook, the more delicious they become.

> 2½ cups dried navy beans or other small white beans
> 1 onion, chopped
> 1 15-ounce can tomato sauce
> ½ cup molasses
> 2 teaspoons stone-ground or Dijon mustard
> 2 tablespoons vinegar
> ½ teaspoon garlic granules or powder
> 1 teaspoon Bakkon yeast* (or torula yeast) (optional)
> 1 teaspoon salt

---

*Important! Bakkon yeast is *not* the same as baking yeast! Bakkon is a brand name for torula yeast, which has a distinctive smoky flavor and is sold in natural food stores.

Rinse the beans thoroughly and soak in 6 cups of water for 6 to 8 hours. Discard the water and place the beans and chopped onion in a pot with enough fresh water to cover the beans with 1 inch of liquid. Bring to a simmer, then cover and cook until the beans are tender, 2 to 3 hours. Add the tomato sauce, molasses, mustard, vinegar, garlic granules, and Bakkon yeast (if using). Cook, loosely covered, over very low heat for 1 to 2 hours. Or, transfer to an ovenproof dish and bake at 350° F for 2 to 3 hours. Add salt to taste.

VARIATION: Place the cooked beans into a crockpot with all the remaining ingredients. Cover and cook on high for 2 to 3 hours.

Per ½ cup: 129 calories; 5 g protein; 27 g carbohydrate; 0 g fat; 164 mg sodium

Red Beans and Rice with Broccoli

Makes about 12 cups

This is a hearty one-dish meal.

1 cup long grain brown rice
4 cups boiling water
½ teaspoon salt
1 onion, chopped
1 celery stalk, thinly sliced
4 garlic cloves, minced
1 red bell pepper, diced
1 teaspoon thyme
¼ teaspoon crushed red pepper
¼ teaspoon black pepper
1 15-ounce can red beans, undrained

2 15-ounce cans black-eyed peas, undrained
1 bunch broccoli
¼ cup Sesame Salt (page 207)

Rinse the rice and add it to a pot with the water and salt. Bring to a simmer, then cover loosely and cook until the rice is tender, about 25 minutes. Pour off excess liquid (this can be saved and used as a broth for cooking if desired).

While the rice cooks, heat ½ cup of water in a large pot and add the onion, celery, and garlic. Cook over medium-high heat, stirring often until the onion is soft, about 5 minutes. Add the bell pepper and seasonings. Cook until the pepper is very soft and all the water has evaporated, about 5 minutes. Stir in the red beans and black-eyed peas, including liquid. Cover and simmer, stirring occasionally, for 20 minutes.

Rinse the broccoli and cut off the stems. Cut or break into bite-sized florets and place on a vegetable steamer. Peel the stems, then slice into rounds and add to the steamer. Steam over boiling water until bright green and tender, about 5 minutes. Toss with Sesame Salt.

Place a serving of rice on a plate or in a large bowl. Top with about ½ cup of beans and about ½ cup of broccoli.

Per 1½ cups: 241 calories; 9 g protein; 44 g carbohydrate; 3 g fat; 399 mg sodium

❊ Hearty Barbecue Beans ❊

Makes 6 cups

1 16-ounce can vegetarian baked beans
1 15-ounce can kidney beans
1 10-ounce package frozen baby lima beans

1 6-ounce can crushed tomatoes
1 cup finely chopped onion
1 tablespoon cider vinegar
1 tablespoon molasses
2 teaspoons stone-ground mustard
1 teaspoon chili powder

Combine all the ingredients in a saucepan and cook at a slow simmer for 25 to 30 minutes.

Per ½ cup: 84 calories; 4 g protein; 16 g carbohydrate; 0 g fat; 221 mg sodium

Curried Potatoes and Chickpeas

Serves 8

Serve with basmati rice or couscous.

2 teaspoons olive oil
1 large onion, chopped
1 tablespoon whole cumin seeds
2 large potatoes
1 cup crushed or ground tomatoes
1 15-ounce can garbanzo beans, including liquid
1 teaspoon turmeric
1 teaspoon coriander
½ teaspoon ginger
¼ teaspoon cayenne
½ teaspoon salt

Heat the oil in a large pot, then add the onion and cook over high heat until soft, about 5 minutes. Add the cumin seeds and continue cooking, stirring often, another 2 to 3 minutes. Scrub the potatoes and cut them into ½-inch cubes. Add them to the onions, along with the tomatoes, garbanzo beans, spices, and salt. Stir in ½ cup of water. Bring to a slow simmer, then cover and cook, stirring occasionally, until the potatoes are tender, about 20 minutes.

Per serving: 142 calories; 4 g protein; 27 g carbohydrate; 2 g fat; 211 mg sodium

❋ Edamame ❋
(Whole Green Soybeans)

Serves 6

Edamame is a Japanese appetizer of green soybeans that are boiled and served in the pod, much like serving peanuts in the shell.

> 1 pound green soybeans (edamame)
> ½ teaspoon salt (optional)

Bring 6 cups of water to a boil in a large pot. Add the soybeans and return to a boil. Cook 10 minutes. Drain well and toss with salt. Shell pods before eating.

Per ½ cup shelled beans: 120 calories; 10 g protein; 8 g carbohydrate; 5 g fat; 10–120 mg sodium

❋ Always Great Brown Rice ❋

Makes 3 cups

As with pasta, cooking rice in extra water ensures a perfect product every time and actually reduces cooking time. Long-grain brown rice gives the lightest, most tender results, while short-grain brown rice is perfect for heartier dishes. Nutritionally, there is very little difference between the two. My personal favorites are the aromatic brown varieties, such as basmati and jasmine, which are sold in natural food stores and many supermarkets.

> 1 cup short- or long-grain brown rice
> 3 cups water
> ½ teaspoon salt (optional)

Rinse the rice in a saucepan of cool water, then drain off as much water as possible. Place the saucepan over medium heat, stirring constantly until the rice dries, 3 to 5 minutes. Add at least 3 cups of water. Cover and simmer until the rice is just tender, about 35 minutes. Pour off excess liquid (this can be saved and used as a broth for soups and stews if desired).

Per ½ cup: 115 calories; 2.5 g protein; 25 g carbohydrate; 1 g fat; 178 mg sodium

❋ Quick Confetti Rice ❋

Makes about 3 cups

This colorful rice pilaf has no added fat, so be sure to cook it in a nonstick pan.

2 cups cooked brown rice (page 270)
½ cup frozen corn
½ cup frozen peas
½ cup diced red bell pepper
½ teaspoon curry powder
¼ cup raisins (optional)
Salt to taste

Heat ¼ cup of water in a large skillet. Add the cooked rice, using a spatula to separate the rice kernels. Add the corn, peas, bell pepper, curry powder, and raisins. Heat thoroughly. Add salt to taste.

Per ½ cup: 109 calories; 2.5 g protein; 24 g carbohydrate; 0 g fat; 112 mg sodium

✖ Ginger Noodles ✖

Serves 6 to 8

These exotic-tasting noodles are surprisingly easy to prepare. Soba noodles are made from buckwheat flour and are sold in natural food stores and Asian markets.

1 package soba noodles (approximately
  8 ounces)
3 tablespoons seasoned rice vinegar
3 tablespoons reduced-sodium soy sauce
2 teaspoons finely chopped fresh ginger
2 garlic cloves, minced
½–1 jalapeño pepper, finely chopped
2 green onions, finely chopped, including tops
¼ cup chopped fresh cilantro (optional)

The Recipes

Cook the noodles in boiling water according to package directions. When tender, drain and rinse. Mix all the remaining ingredients, then pour over the noodles and toss to mix.

Per serving: 103 calories; 4 g protein; 21 g carbohydrate; 0.5 g fat; 344 mg sodium

Bulgur

Makes 2½ cups

Bulgur is a delicious grain that is easily prepared. It is made from whole wheat kernels that have been cracked and toasted, giving it a wonderful, nutty flavor. Serve it plain or use it in pilafs and salads. Bulgur is sold in natural food stores and in some supermarkets. Check the bulk food section.

1 cup uncooked bulgur
½ teaspoon salt
2 cups boiling water

Mix the bulgur and salt in a large bowl, then pour the boiling water over it. Cover and let stand until tender, about 25 minutes.

VARIATION: Bring the water to a boil, then add the salt and bulgur. Reduce heat to a simmer, then cover and cook until the bulgur is tender, about 15 minutes.

Per ½ cup: 112 calories; 4 g protein; 24 g carbohydrate; 0 g fat; 213 mg sodium

❈ Couscous ❈

Makes 3 cups

Though it looks like a grain, couscous is the world's smallest pasta. It takes only minutes to prepare and makes a delicious side dish or salad ingredient. Whole wheat couscous, which contains fiber and more vitamins and minerals than refined couscous, is sold in natural food stores and some supermarkets.

> ½ teaspoon salt
> 1 cup whole wheat couscous

Bring 1½ cups of water to a boil, then stir in the salt and couscous. Remove from the heat and cover. Let stand 10 to 15 minutes, then fluff with a fork and serve.

Per ½ cup: 91 calories; 3 g protein; 20 g carbohydrate; 0 g fat; 93 mg sodium

❈ Polenta ❈

Makes 4 cups

It is no surprise that polenta, not pasta, is the staple grain of northern Italy. It's delicious, easy to prepare, and tremendously versatile. Straight out of the pan it is soft, like cream of wheat. It may be served as a breakfast cereal topped with fruit and soy milk or as a hearty entrée topped with a savory sauce. When chilled, it becomes firm and sliceable, perfect for grilling or sautéeing (Breakfast Scramble, page 183). It may also be used as a pizza crust (Polenta Pizza, page 276).

> 5 cups water
> 1 cup polenta

The Recipes

1 teaspoon salt
1 teaspoon each: thyme, oregano (optional)

Measure water into a large pot, then whisk in the polenta, salt, and herbs (if using). Simmer over medium heat, stirring often, until very thick, about 25 minutes. Serve hot or transfer to a 9 × 13-inch baking dish and chill until firm.

Per ½ cup: 62 calories; 1 g protein; 14 g carbohydrate; 0 g fat; 267 mg sodium

❈ Zucchini Corn Fritters ❈

Makes sixteen 3-inch fritters

These "fritters" are not deep fried, but actually cooked like pancakes.

1⅓ cups soy milk
1 tablespoon cider vinegar
1 cup cornmeal
¼ cup unbleached flour
½ teaspoon baking powder
½ teaspoon baking soda
½ teaspoon salt
1 medium zucchini (about 1 cup grated or finely chopped)
1 cup corn (fresh, frozen, or canned)

Combine the soy milk and vinegar and set aside. In a mixing bowl, combine the cornmeal, flour, baking powder, baking soda, and salt.

Chop or grate the zucchini (a food processor makes it easy), then add it to the cornmeal mixture along with the soy milk mixture and corn. Stir to mix.

Turn Off the Fat Genes

Mist a preheated nonstick skillet with vegetable oil spray. Pour small amounts of batter into the pan and cook the first side until the edges are dry, about 2 minutes. Carefully turn the fritters and cook the second side until browned, about 1 minute.

Per fritter: 55 calories; 2 g protein; 11 g carbohydrate; 0 g fat; 101 mg sodium

❋ Quinoa ❋

Makes 3 cups

Quinoa (pronounced "keen-wah") comes from the high plains of the Andes Mountains, where it is nicknamed "the mother grain" for its life-giving properties. The National Academy of Sciences has called quinoa "one of the best sources of protein in the vegetable kingdom" because of its excellent amino acid pattern. Quinoa cooks quickly, and as it cooks the germ unfolds like a little tail. It has a light, fluffy texture, and may be eaten plain, used as a pilaf, or as an addition to soups and stews. Quinoa is coated with a bitter-tasting substance called saponin that repels insects and birds and protects it from ultraviolet radiation. Be sure to wash quinoa thoroughly before cooking to remove this bitter coating. The easiest way to wash quinoa is to place it in a strainer and rinse it with cool water until the water runs clear.

1 cup quinoa
2 cups water

Rinse quinoa thoroughly in a fine sieve. Bring the water to a boil, then add the quinoa and bring to a boil again. Reduce to a simmer, then cover loosely and cook 15 minutes.

Per ½ cup: 118 calories; 5 g protein; 20 g carbohydrate; 2 g fat; 3 mg sodium

The Recipes

❈ Polenta Pizza ❈

Serves 12

1 cup polenta
5 cups water
½ teaspoon salt
½ teaspoon each: thyme, oregano
10 sundried tomato halves
½ cup boiling water
1 large onion, finely chopped
6 large garlic cloves
8 ounces chopped mushrooms
12 ounces water-packed roasted red peppers,
 finely chopped
1 teaspoon basil
1 pound firm reduced-fat tofu, squeezed
1½ tablespoons white miso
2 teaspoons basil
1 tablespoon tahini (sesame seed butter)
3–4 garlic cloves, minced
2 10-ounce packages frozen chopped spinach,
 thawed
2 teaspoons balsamic vinegar
1 tablespoon reduced-sodium soy sauce
⅛ teaspoon black pepper
1 cup tomato sauce

Whisk polenta into water, then add salt and herbs. Simmer over medium heat, stirring often, until very thick, about 25 minutes. Spread on a large baking sheet. Cool.

Soak the sundried tomatoes in boiling water until soft, about 20 minutes. Heat ½ cup of water and cook the onion, garlic, and mushrooms until the onion is soft and the mushrooms are brown. Chop the softened tomatoes and add them along with the roasted red peppers and 1 teaspoon basil. Simmer 5 minutes.

Place the tofu in a clean dish towel and squeeze it to remove some of the water. Transfer the squeezed tofu to a food processor with the miso, 2 teaspoons basil, and tahini. Process until completely smooth.

Heat ¼ cup of water in a large skillet, then add the garlic. Cook 1 minute; then add the spinach, vinegar, soy sauce, and pepper. Cook, stirring often, until all liquid evaporates.

Preheat oven to 350° F. To assemble pizza: spread tofu mixture evenly over polenta. Top with spinach, then with tomato sauce. Bake in preheated oven for 20 minutes.

Per 1/10 of pizza: 165 calories; 9 g protein; 26 g carbohydrate; 2 g fat; 326 mg sodium

Eggplant Manicotti

Makes about 12 manicotti

1 large eggplant
1 medium onion, chopped
1 10-ounce package frozen chopped spinach, thawed and squeezed dry
1 teaspoon basil
¼ teaspoon oregano
½ teaspoon onion powder
½ teaspoon garlic powder
⅛ teaspoon nutmeg
2 tablespoons whole wheat flour
2 cups Simple Marinara (page 202)

Slice the eggplant lengthwise into 12 ¼-inch-thick slices. Spray a nonstick skillet with cooking spray and cook the first side until it is slightly browned. Then turn and cook the second side until the eggplant is tender when pierced with a fork. Set aside.

Heat ¼ cup of water in a large nonstick skillet and cook the onion over medium-high heat until all the liquid has evaporated. Add 2 tablespoons of water and stir to loosen any stuck bits of onion. Continue cooking until all the liquid evaporates again, then stir in 2 more tablespoons of water. When the liquid has evaporated, add the spinach and seasonings. Stir in the flour and cook 2 to 3 minutes longer. Set aside to cool.

Preheat oven to 350° F. Place a spoonful of the spinach mixture across the center of each slice of eggplant; then, beginning with the narrow end, roll the eggplant around the filling. Arrange in a baking dish, seam side down. Top with Simple Marinara. Cover and bake for 20 minutes.

Per manicotti (with marinara): 42 calories; 1 g protein; 9 g carbohydrate; 0 g fat; 40 mg sodium

Simple Pasta Supper

Makes about 6 cups

8 ounces pasta spirals
1 tablespoon olive oil
1 onion chopped
¼ cup chopped garlic
1 cup tomato juice or V-8 juice
1 15-ounce can red kidney beans, including liquid
2 cups finely chopped fresh kale
½ cup chopped fresh basil
¼ teaspoon salt

Cook the pasta until just tender. Transfer to a collander. Rinse and drain. Set aside.

Heat the oil in a large skillet or pot. Add the onion and garlic and cook over high heat, stirring often until the onion begins to brown, about 6 minutes. Stir in ¼ cup of water, scraping the pan to remove any bits of onion. Add the tomato juice, kidney beans with their liquid, kale, and basil. Stir to mix, then cover and simmer, stirring occasionally until the kale is tender, about 5 minutes. Stir in the cooked pasta. Add salt to taste before serving.

Per cup: 126 calories; 5 g protein; 22 g carbohydrate; 2 g fat; 372 mg sodium

❋ Black Bean Pueblo Pie ❋

Makes one 9 × 13-inch casserole

This mouthwatering casserole combines black bean chili with corn tortillas and two spicy sauces. The recipe has several steps, but the results are well worth the effort.

> 4 cups cooked black beans, or 2 15-ounce cans
> 1 15-ounce can crushed tomatoes
> 2 teaspoons paprika
> 2 tablespoons chili powder
> 2 teaspoons onion powder
> 1 teaspoon garlic powder
> 1 large onion, chopped
> 1 tablespoon minced garlic (about 4 large cloves)
> 1 28-ounce can crushed tomatoes
> 2½ teaspoons cumin

1 15-ounce can garbanzo beans, drained
½ cup water-packed roasted red pepper (about
   2 peppers)
2 garlic cloves, peeled
1 tablespoon tahini (sesame seed butter)
3 tablespoons lemon juice
12 corn tortillas, torn in half
1 cup chopped green onions

Combine the black beans, including the liquid, with the tomatoes, paprika, 2 teaspoons of chili powder, onion powder, and garlic powder in a pot. Simmer, stirring frequently, for 25 minutes.

To make the tomato sauce, heat ½ cup of water in a large skillet, then add the onion and garlic. Cook over high heat, stirring occasionally, for 5 minutes. Stir in the tomatoes, 4 teaspoons of chili powder, and 2 teaspoons of cumin. Cover and simmer over medium heat for 5 minutes.

Process the garbanzo beans, roasted peppers, garlic, tahini, lemon juice, and the remaining ½ teaspoon of cumin in a food processor or blender until very smooth.

Preheat oven to 350°F. Spread ½ to 1 cup of the tomato sauce in the bottom of a 9 × 13-inch (or larger) baking dish. Cover with a layer of tortillas, then spread with half of the garbanzo mixture, using your fingers to hold the tortillas in place. Top with half of the black beans and sprinkle with green onions. Spread with half of the tomato sauce. Repeat the layers, ending with tomato sauce. Bake for 25 minutes.

Per ¹/₁₂ of casserole: 198 calories; 8 g protein; 36 g carbohydrate; 2 g fat; 267 mg sodium

Turn Off the Fat Genes

�֎ Potato Enchiladas ✖

Makes 12 enchiladas

SAUCE:

1 onion, chopped
2 garlic cloves, minced
½ teaspoon cumin
1 12-ounce jar roasted red peppers
1½ cups salsa or picante sauce (use your
  favorite brand or try Pace Picante)

FILLING:

1 15-ounce can garbanzo beans, drained
½ cup roasted red pepper, drained
1 tablespoon tahini
3 tablespoons lemon juice
1 garlic clove
¼ teaspoon cumin
12 corn tortillas
4 cups cooked, diced potatoes
1–2 cups chopped green onions
1–2 cups salsa or picante sauce (use your
  favorite brand or try Pace Picante)

To make the sauce, heat ½ cup of water in a pot and add the onion, garlic, and cumin. Cook over high heat, stirring occasionally, until the onion is soft and translucent, about 5 minutes. Pour the roasted red peppers, including any liquid, into a blender and blend until completely smooth. Add to the onions along with 1½ cups of salsa or picante sauce and simmer 5 minutes.

For the filling, place the drained garbanzo beans in a food processor. Add ½ cup of drained roasted red peppers along with the tahini, lemon juice, garlic, and cumin. Process until completely smooth, 2 to 3 minutes. Preheat oven to 350°F.

Warm the tortillas. This can be done in several ways: cut a slit in a bag to allow steam to escape, then place the entire bag of tortillas into a microwave and heat for 30 seconds or until warm and soft; or wrap the tortillas in foil and heat them in a 350°F oven for about 20 minutes; or place a single layer of tortillas on a vegetable steamer and steam until soft, about 10 seconds. Repeat with remaining tortillas.

To assemble the enchiladas, spread a warmed tortilla with about 2 tablespoons of the garbanzo mixture and place it in a 9 × 13-inch baking dish that has been lightly oil-sprayed (or use a nonstick dish). Arrange about 2 tablespoons of the potatoes in a line across the center of the tortilla. Sprinkle with about 1 tablespoon of green onions and 1 to 2 tablespoons of the salsa or picante sauce. Roll the tortilla around the filling and arrange it in the dish, seam side down. Repeat with the remaining tortillas. Cover with foil and bake 20 minutes. Uncover and top with sauce before serving.

Per enchilada: 210 calories; 6 g protein; 40 g carbohydrate; 2 g fat; 259 mg sodium

❋ Tamale Pie ❋

Makes one 9 × 13 casserole

5 cups water
1 cup polenta
1 teaspoon salt
1 large yellow onion, chopped
3 large garlic cloves, minced
1 red or green bell pepper, seeded and diced
½ cup chopped fresh cilantro (optional)

1 12-ounce package Yves Veggie Ground Round
   (or similar product)
1 tablespoon chili powder
1 teaspoon cumin
1 teaspoon coriander (optional)
1 28-ounce can crushed tomatoes
1 15-ounce can vegetarian chili beans, including
   liquid

Measure water into a large pot, then whisk in the polenta and salt. Simmer over medium heat, stirring often, until very thick, about 25 minutes. Set aside.

Heat ½ cup of water in a large skillet or pot and add the onion, garlic, bell pepper, and cilantro, if using. Cook until the onion is soft and transluscent, about 5 minutes, stirring occasionally.

Add the Veggie Ground Round, chili powder, cumin, and coriander (if using) and cook over medium heat, stirring often, for 3 minutes. Add a bit more water if needed to prevent sticking.

Stir in the crushed tomatoes and chili beans. Cover and simmer 10 minutes. Transfer to a 9 × 13-inch baking dish. Spread the cooked polenta evenly over the top. Bake at 350°F for 20 minutes.

Per ¹/₁₀ of casserole: 148 calories; 9 g protein; 22 g carbohydrate; 3 g fat; 507 mg sodium

❋ Mexican Skillet Pie ❋

Makes about 12 cups

1 15-ounce can garbanzo beans, drained
½ cup water-packed roasted red peppers
   (about 2 peppers)

2 garlic cloves, peeled

1 tablespoon tahini (sesame seed butter)

3 tablespoons lemon juice

½ teaspoon cumin

1 onion, chopped (optional)

4 garlic cloves, minced (optional)

1 28-ounce can crushed tomatoes

12 corn tortillas

2 15-ounce cans vegetarian chili beans

1 cup roasted red peppers (optional)

1 15-ounce can corn kernels, drained

3 green onions, chopped (optional)

Combine the garbanzo beans, peppers, garlic, tahini, lemon juice, and cumin in a food processor or blender and process until completely smooth. Set aside. If using the onion and garlic, heat ½ cup of water in a large skillet, then add the onion and garlic and cook over high heat for 5 minutes, stirring occasionally. Remove from heat and add the crushed tomatoes, spreading them evenly over the bottom of the pan.

Arrange the tortillas over the tomatoes (there will be several layers).

Spread the chili beans evenly over the tortillas and top with the roasted red peppers (if using), the drained corn, and an even layer of the garbanzo mixture.

Cover the skillet and cook over medium heat until hot and steamy, 15 to 20 minutes. Sprinkle with chopped green onions, if desired, and serve.

Per ¹/₁₂ of casserole: 242 calories; 9 g protein; 45 g carbohydrate; 3 g fat; 537 mg sodium

Turn Off the Fat Genes

❋ Vegetable Curry ❋

Makes about 8 cups

Serve with cooked brown rice (page 270) and Apple Chutney (page 207).

2 tablespoons reduced-sodium soy sauce

1 large onion, sliced

1 large sweet potato or yam, peeled and diced (about 2 cups)

1 large carrot, thinly sliced

1 celery stalk, thinly sliced

1 red or green bell pepper, diced

1 15-ounce can crushed tomatoes

1 15-ounce can garbanzo beans, including liquid

1 teaspoon mustard seeds

½ teaspoon turmeric

½ teaspoon cumin

½ teaspoon coriander

½ teaspoon ginger

⅛ teaspoon cayenne

Cooked brown rice for serving

Apple chutney for serving

Heat ½ cup of water and the soy sauce in a large pot. Add the onion and diced sweet potato. Cook over high heat, stirring occasionally, for 5 minutes. Stir in the carrot, celery, and bell pepper. Cover and continue cooking another 3 minutes, stirring occasionally. Add the tomatoes, garbanzo beans and liquid, and spices. Stir to mix; cover and simmer over medium heat, stirring occasionally. Add a small amount of water if the mixture begins to stick. Cook until the vegetables are tender, about 10 minutes.

Per cup: 134 calories; 4 g protein; 28 g carbohydrate; 0.5 g fat; 407 mg sodium

❋ No-Meat Loaf ❋

Makes 12 slices

This savory loaf is delicious with Mashed Potatoes and Gravy (page 252) or as a sandwich filling. The vegetables need to be finely chopped, which can be easily accomplished with a food processor.

1 cup bulgur wheat
1½ cups boiling water
1 small onion, finely chopped
1 medium carrot, shredded or finely chopped
2 stalks celery, finely chopped
1 pound mushrooms, finely chopped
½ cup finely chopped walnuts
⅓ cup potato flour
½ teaspoon each: marjoram, garlic powder, salt
¼ teaspoon each: thyme, sage, black pepper
3 tablespoons ketchup or barbecue sauce
1 tablespoon reduced-sodium soy sauce
4 teaspoons stone-ground mustard

Additional ketchup or barbecue sauce for topping

Place the bulgur in a large bowl and pour the boiling water over it. Soak until the bulgur is tender and most of the water is absorbed, about 15 minutes. Heat 2 tablespoons of water in a nonstick skillet and add the onion, carrot, and celery. Cook over medium-high heat for 3 minutes, stirring often. Stir in the mushrooms and continue cooking, stirring occasionally, until the vegetables are soft and the mushrooms are brown, about 5 minutes.

Preheat oven to 350° F. Drain any excess water off the bulgur. Add the vegetables along with the remaining ingredients and stir for 1 to 2 minutes, until the mixture holds together. Pat into an oil-sprayed 5 × 9-inch loaf pan and top with barbecue sauce or ketchup. Bake for 60 minutes. Let stand for 10 minutes before serving.

Per slice: 109 calories; 3 g protein; 17 g carbohydrate; 3 g fat; 223 mg sodium

❋ Stuffed Winter Squash ❋

Serves 6

Don't let the name fool you: "winter" squash are available most of the year and make a delicious meal any time.

3 medium-sized winter squash (acorn, kabocha, sweet dumpling, or delicata)
3 tablespoons reduced-sodium soy sauce
1 medium onion, chopped
2 garlic cloves, minced
2 cups mushrooms, sliced
1 cup sliced celery
¼ cup finely chopped parsley
4 cups cubed whole wheat bread
½ cup dried apricots, chopped
½ teaspoon each: sage, marjoram, thyme
¼ teaspoon black pepper
2 cups apricot nectar
¼ teaspoon ginger
¼ teaspoon coriander
¼ teaspoon cinnamon
2 tablespoons maple syrup
2 teaspoons cornstarch

Cut the squash in half and scoop out the seeds. Place in a vegetable steamer and steam until tender when pierced with a fork. This should take 20 to 30 minutes. Heat ½ cup of water in a large pot and add 1 tablespoon of the soy sauce, the onion, garlic, mushrooms, and celery. Cover and cook over medium heat, stirring occasionally, until the onion is soft, about 5 minutes. Stir in the parsley, bread cubes, apricots, sage, marjoram, thyme, and black pepper. The mixture should be moist enough to hold together, but not wet. If it is too dry, add a small amount of water or vegetable stock.

Preheat oven to 350° F. Divide the stuffing mixture evenly among the squash halves and bake 20 minutes. Meanwhile, mix the apricot nectar with the ginger, coriander, cinnamon, remaining soy sauce, maple syrup, and cornstarch. Bring to a simmer, stirring constantly, and cook until clear and slightly thickened, about 2 minutes. Remove from heat and set aside.

Use a spoon to make a depression in the top of the stuffing on each squash. Fill with a spoonful or two of the apricot sauce. Serve remaining sauce on the side.

Per serving: 283 calories; 5 g protein; 49 g carbohydrate; 1 g fat; 304 mg sodium

Portabellos with Collards and Cannelini Beans

Makes eight 1-cup servings

Thick slices of pan-grilled portabello mushrooms make a hearty meal with brown and wild rice.

3½ cups water
½ teaspoon salt
½ cup wild rice

1 cup long-grain brown rice

1 teaspoon olive oil

2 tablespoons reduced-sodium soy sauce

1 teaspoon thyme

½ teaspoon oregano

¼ teaspoon black pepper

1 onion, chopped

3 garlic cloves, minced

2 large Portabello mushrooms
(about ½ pound), thickly sliced

4 cups chopped collard greens
(about ½ pound)

1 15-ounce can cannelini beans, drained

Bring water to a boil, then add the salt and wild rice. Cover and simmer 15 minutes. Add the brown rice, then cover and simmer until the rice is tender, about 45 minutes. Drain off any excess water. While the rice is cooking, prepare the mushrooms. Mix the oil, 1 tablespoon of the soy sauce, the thyme, oregano, and black pepper with ¼ cup of water in a large nonstick skillet. Add the onion, garlic, and mushrooms. Cover and cook over medium-high heat, stirring occasionally, until the onion is soft and the mushrooms are browned, about 5 minutes. Reduce the heat to medium and add ¼ cup of water and the collards. Sprinkle with the remaining 1 tablespoon of soy sauce. Cover and cook until the collards are just tender, 6 to 8 minutes, then stir in the drained beans. Cover and cook 1 to 2 more minutes. Serve over the cooked rice.

Per cup: 199 calories; 7 g protein; 38 g carbohydrate; 1 g fat; 361 mg sodium

The Recipes

Tofu Vegetable Hash

Makes about 8 cups

2 russet potatoes, scrubbed and diced
1 tablespoon extra virgin olive oil
1 large onion, sliced
2½ cups sliced mushrooms
1 pound firm reduced-fat tofu, rinsed and cut into ½-inch cubes
1 bunch fresh basil, stemmed and chopped
2 tomatoes, diced
2 cups broccoli florets
2 bunches baby bok choy, thinly sliced
3 tablespoons reduced-sodium soy sauce
¼ teaspoon black pepper

Steam the potatoes until they are just tender when pierced with a fork. Set aside. While the potatoes are cooking, prepare the rest of the vegetables as directed. Warm the oil in a large nonstick skillet, then add the onion and mushrooms. Cook over high heat, stirring often, until the onion is golden and the mushrooms are brown, about 5 minutes. Add the tofu, basil, and ¼ cup of water. Cook 3 minutes over high heat, stirring often, adding small amounts of water if needed to prevent sticking. Reduce heat to medium and add the remaining ingredients with ¼ cup of water. Stir gently to mix, then cover and cook until the broccoli is just tender, about 5 minutes. Serve hot.

Per cup: 155 calories; 9 g protein; 22 g carbohydrate; 3 g fat; 248 mg sodium

❋ Lasagne Roll-ups ❋

Makes about 16 roll-ups

In this recipe, a creamy spinach filling is rolled up in cooked lasagne noodles.

½ cup red wine, vegetable stock, or water
1 small onion, chopped
3 garlic cloves, pressed or minced
1½ cups sliced mushrooms (about ½ pound)
¼ cup chopped parsley
1 15-ounce can crushed or ground tomatoes
1½ tablespoons apple juice concentrate
½ cup water
½ teaspoon basil
½ teaspoon oregano
¼ teaspoon fennel seeds (optional)
¼ teaspoon black pepper
8–10 wide lasagne noodles

1 garlic clove
1 pound firm reduced-fat tofu
2 10-ounce packages frozen chopped spinach,
    thawed and squeezed
½ cup finely chopped parsley
1 teaspoon basil
½ teaspoon oregano
½ teaspoon thyme
½ teaspoon nutmeg
½ teaspoon salt
¼ teaspoon black pepper

To make the sauce, heat the wine, stock, or water in a large pot. Add the onion and garlic and cook until the onion is soft, about 5 minutes. Add the mushrooms and parsley. Lower the heat slightly, cover, and cook until the mushrooms are soft, about 5 minutes. Stir in the tomatoes, apple juice concentrate, water, basil, oregano, fennel seeds, and black pepper. Cover and simmer 15 minutes.

Cook the noodles in boiling water until they are just tender. Pour into a colander and rinse in cold water. Set aside.

To prepare the filling, finely chop the garlic in a food processor, then add the tofu and process until completely smooth. Mix in the spinach, parsley, basil, oregano, thyme, nutmeg, salt, and black pepper.

Preheat oven to 350°F. Spread the sauce evenly over the bottom of a 9 × 13-inch baking dish. Cut a noodle in half so that it is about 5 inches long and spread it evenly with about ¼ cup of the filling mixture. Place it in the baking dish (seam side down or standing on end) and repeat with the remaining noodles. Cover and bake in the preheated oven until very hot, about 20 minutes.

Per roll-up: 164 calories; 9 g protein; 27 g carbohydrate; 2 g fat; 99 mg sodium

DESSERTS

❈ Strawberry Freeze ❈

Makes about 2 cups

Frozen fruit is the secret to making this thick, rich-tasting dessert. Frozen strawberries, without added sugar, are sold in most markets. To freeze bananas, peel and break them into chunks, then pack them loosely in an airtight container. Frozen strawberries will keep about six months, and frozen bananas for two to three months.

1 cup frozen strawberries
1 cup frozen banana chunks
½–1 cup unsweetened apple juice

Place all the ingredients into a blender and process on high until smooth, stopping the blender to move the unblended fruit to the center with a spatula.

Per cup: 116 calories; 1 g protein; 27 g carbohydrate; 0 g fat; 8 mg sodium

❋ Fresh Peach Freeze ❋

Makes about 5 cups

This versatile freeze makes a healthful breakfast, a delicious dessert, or a satisfying snack any time of day. You can buy frozen peach slices at most food stores, or you can freeze your own during the summer when peaches are ripe and bursting with flavor. Select a freestone variety so the pits can be easily removed, then cut the peaches (or nectarines) into ¼-inch-thick slices. Place into zipper bags to freeze. By freezing 4 cups in each bag, you won't have to measure when you make this recipe. Just take a bag from the freezer, flex it a bit to separate the peach slices, then add the slices to the blender along with the other ingredients.

> 4 cups frozen peach or nectarine slices
> ½ cup apple juice concentrate
> 2 cups vanilla soy milk or vanilla rice milk

Combine all the ingredients in a blender. Hold the lid on firmly and blend at high speed until completely smooth, stopping the blender to stir any unblended fruit into the center.

Per cup: 135 calories; 3 g protein; 28 g carbohydrate; 1 g fat; 48 mg sodium

❈ Chocolate Banana Smoothie ❈

Makes about 2 cups

2 cups frozen banana chunks
½ cup chocolate soy milk or rice milk
2 tablespoons maple syrup

Combine all ingredients in a blender and blend until smooth, adding a bit more soy milk if necessary.

Per cup: 183 calories; 3 g protein; 39 g carbohydrate; 1.5 g fat; 45 mg sodium

❈ Fresh Peach Crisp ❈

Makes one 9 × 9-inch crisp

This is a perfect dessert when summer peaches (or nectarines) are at their peak. Frozen peaches may be substituted for fresh during other times of year.

5–6 fresh peaches
1 tablespoon cornstarch
1 tablespoon sugar
1 tablespoon lemon juice
1¼ cups Grape-Nuts cereal
1½ cups rolled oats
½ cup Corn Butter (page 200)
½ cup maple syrup
½ teaspoon cinnamon
1 teaspoon vanilla

Peel and slice enough peaches to make 6 cups. Toss with cornstarch and sugar, then spread in a 9 × 9-inch baking dish and

Turn Off the Fat Genes

drizzle with the lemon juice. Preheat oven to 350° F. Mix the remaining ingredients and spread evenly over the peaches. Bake until topping is golden brown, about 25 minutes.

Per 1/9 of crisp: 213 calories; 5 g protein; 46 g carbohydrate; 1 g fat; 141 mg sodium

Apple Crisp

Makes one 9 × 9-inch crisp

Use pippins, Granny Smiths, or similar tart green apples for this dessert.

4 large green apples, peeled and sliced
1 teaspoon cinnamon
1 tablespoon lemon juice
½ cup raisins
1 cup Grape-Nuts cereal
1 cup rolled oats
½ cup maple syrup
⅔ cup apple juice
1 tablespoon cornstarch

Preheat oven to 350° F. Toss the apple slices with ½ teaspoon of the cinnamon. Arrange in a 9 × 9-inch baking dish and sprinkle with lemon juice and raisins. Mix the Grape-Nuts, rolled oats, and remaining cinnamon. Stir in the maple syrup, then spread evenly over the apples. Mix the apple juice and cornstarch, then pour evenly over other ingredients. Bake until the apples are tender when pierced with a knife, 35 to 50 minutes.

Per 1/9 of crisp: 196 calories; 3 g protein; 44 g carbohydrate; 1 g fat; 89 mg sodium

Fruit Gel

Makes 8 cups

This is an all-natural alternative to Jell-O! Agar powder and kudzu are plant-based thickeners available in natural food stores.

3 cups strawberries, fresh or frozen
1½ cups apple juice concentrate
1 cup water
1 teaspoon agar powder
2 tablespoons kudzu powder
4 cups blueberries, fresh or frozen

Chop the strawberries by hand or in a food processor and place them in a pan with the apple juice concentrate, water, agar, and kudzu. Stir to mix. Bring to a simmer and cook 3 minutes, stirring often. Remove from heat and chill completely. Fold in blueberries and transfer to serving dishes.

Per cup: 145 calories; 1 g protein; 34 g carbohydrate; 0 g fat; 19 mg sodium

Gingerbread

Makes one 9 × 9-inch cake

This gingerbread is delicious topped with Pineapple Apricot Sauce (page 204). You can also use spreadable fruit, thinned with a bit of water, as a topping.

½ cup raisins
½ cup pitted dates, chopped
¾ cup sugar

½ teaspoon salt

2 teaspoons cinnamon

1 teaspoon ginger

¾ teaspoon nutmeg

¼ teaspoon cloves

1¾ cups water

2 cups whole wheat pastry flour

1 teaspoon baking soda

1 teaspoon sodium-free baking powder

Combine the raisins, dates, sugar, salt, spices, and water in a large saucepan and bring to a boil. Boil for 2 minutes, then remove from heat and cool completely (this is important). When cool, preheat oven to 350° F. Stir the flour, baking soda, and baking powder together. Add to the cooled fruit mixture and stir to mix. Spread into a 9 × 9-inch pan that has been sprayed with a nonstick spray and bake for 30 minutes, or until a toothpick inserted into the center comes out clean.

Per 3 × 3-inch slice: 207 calories; 4 g protein; 48 g carbohydrate; 0 g fat; 216 mg sodium

❋ Butterscotch Pudding ❋

Makes about 3 cups

This butterscotch pudding is sweetened with maple syrup and includes cooked yam for color, body, and nutrition.

2 cups soy milk

1 cup cooked, peeled yam

5 tablespoons maple syrup

2 tablespoons cornstarch

1 tablespoon potato flour

The Recipes

½ teaspoon butterscotch extract
¼ teaspoon salt

Combine the soy milk, cooked yam, and maple syrup in a blender and process until smooth. With the blender running, add the remaining ingredients. Transfer to a saucepan and heat, stirring constantly, until the mixture bubbles and thickens. Remove from heat and transfer to individual serving dishes if desired. Cool before serving.

Per ½ cup: 109 calories; 2 g protein; 23 g carbohydrate; 1 g fat; 133 mg sodium

❊ Chocolate Pudding ❊

Makes about 2 cups

This is delicious, old-fashioned chocolate pudding.

2 cups soy milk
3 tablespoons cocoa
3 tablespoons cornstarch
1 tablespoon potato flour
½ cup sugar
1 teaspoon vanilla
⅛ teaspoon salt

Blend or whisk all the ingredients together until smooth. Cook over medium heat, stirring constantly, until the pudding comes to a boil. Continue cooking 30 seconds, stirring constantly. Pour into individual serving dishes and chill.

Per ½ cup: 178 calories; 4 g protein; 35 g carbohydrate; 2 g fat; 115 mg sodium

✻ Carob Mint Pudding ✻

Makes about 2 cups

2 cups soy milk
3 tablespoons carob powder
3 tablespoons cornstarch
⅓ cup sugar
⅛ teaspoon salt
1 tablespoon potato flour
⅛ teaspoon peppermint extract

Blend or whisk all the ingredients together until smooth. Cook over medium heat, stirring constantly, until the pudding comes to a boil. Continue cooking 30 seconds, stirring constantly. Pour into individual serving dishes and chill.

Per ½ cup: 149 calories; 3 g protein; 30 g carbohydrate; 1.5 g fat; 118 mg sodium

✻ Fresh Lemon Curd ✻

Makes about 2 cups

2 cups soy milk
½ cup sugar
3 tablespoons cornstarch
2–3 tablespoons lemon juice
2 teaspoons lemon zest
⅛ teaspoon salt
1 tablespoon potato flour

Combine all the ingredients except the potato flour in a blender and process until smooth. With the blender running, add the potato flour. Transfer to a saucepan and heat, stirring constantly, until the mixture boils. Continue cooking for 30 seconds, stirring constantly. Remove from heat and transfer to individual serving dishes. Cool before serving.

Per ½ cup: 162 calories; 3 g protein; 34 g carbohydrate; 1 g fat; 120 mg sodium

❈ Tapioca Pudding ❈

Makes about 2 cups

2 cups soy milk
¼ cup sugar
1 tablespoon potato flour
⅛ teaspoon salt
3 tablespoons instant tapioca
1 teaspoon vanilla

Blend the soy milk, sugar, potato flour, and salt until smooth. Transfer to a saucepan and stir in the tapioca. Let stand 5 minutes, then bring the mixture to a boil over medium heat, stirring constantly. Remove from heat and stir in the vanilla. Transfer to individual serving dishes. Cool before serving.

Per ½ cup: 115 calories; 3 g protein; 22 g carbohydrate; 1.5 g fat; 117 mg sodium

❈ Apricot Tapioca ❈

Makes about 2 cups

2 cups apricot nectar
3 tablespoons sugar

Turn Off the Fat Genes

½ lemon, juiced (1½ tablespoons)
1 tablespoon potato flour
3 tablespoons instant tapioca
¼ teaspoon almond extract

Blend the apricot nectar, sugar, lemon juice, and potato flour until smooth. Transfer to a saucepan, stir in the tapioca, and let stand 5 minutes. Bring the mixture to a full boil over medium heat, stirring constantly. Remove from heat and stir in the almond extract. Transfer to individual serving dishes if desired. Cool before serving.

Per ½ cup: 115 calories; 3 g protein; 22 g carbohydrate; 1.5 g fat; 117 mg sodium

Ginger Peachy Bread Pudding

Serves 9

1 28-ounce can sliced peaches, packed in juice
1 tablespoon cornstarch
6 cups cubed whole-grain bread (about 8 slices)
1¾ cups soy milk
⅓ cup apple juice concentrate
¾ cup golden raisins
½ teaspoon ginger
½ teaspoon cinnamon
¼ teaspoon nutmeg
¼ teaspoon salt
1 teaspoon vanilla

Drain the liquid from the peaches into a large mixing bowl and add the cornstarch. Stir to dissolve any lumps, then add the bread cubes, soy milk, apple juice concentrate, raisins, ginger, cinnamon, nutmeg, salt, and vanilla. Mix well. Chop the peaches and stir them into the mixture. Spread in an oil-sprayed 9 × 9-inch baking dish. Preheat oven to 350° F. Bake for 35 minutes. Serve warm or cooled.

Per serving: 252 calories; 3 g protein; 57 g carbohydrate; 1 g fat; 185 mg sodium

❋ Poached Pears with ❋ Butterscotch Sauce

Makes 8 pear halves

Poached pears are attractive, delicious, and deceptively easy to prepare.

> 4 large ripe pears, any kind
> 1 cup apple juice concentrate
> 1 cup white wine or water
> ½ teaspoon cinnamon
> 1 cup soy milk
> 3 tablespoons maple syrup
> ⅛ teaspoon salt
> ½ teaspoon butterscotch flavor*
> 1½ tablespoons rice flour
> 2 tablespoons potato flour
> ½ cup Corn Butter (page 200)

Peel the pears, cut them in half, and remove the cores. Place the pears in a pot with the apple juice concentrate, wine or water, and cinnamon. Cover and simmer until tender when pierced with a fork, 15 to 20 minutes. Remove the pears from the pan

---

*Frontier Naturals has a line of natural flavorings that is sold in natural food stores.

Turn Off the Fat Genes

and place in serving dishes. Boil the liquid, uncovered, until it is quite thick, about 5 minutes. Pour it over the pears.

To make the butterscotch sauce, pour the soy milk into a blender. Add the remaining ingredients and blend until completely smooth. Top each pear half with a generous spoonful of sauce.

Per pear half: 157 calories; 2 g protein; 35 g carbohydrate; 1 g fat; 87 mg sodium

✳ Gingerbread Cookies ✳

Makes about 48 cookies

The secret of these wonderful, crisp cookies is rolling the dough very, very thin. The easiest way to prepare the dough is with a heavy-duty mixer.

> ½ cup sugar
> 1 teaspoon powdered ginger
> 1 teaspoon cinnamon
> 1½ teaspoons baking soda
> ¼ teaspoon salt
> ⅓ cup molasses
> ⅓ cup soy milk
> 2¼ cups whole wheat pastry flour

Mix the sugar, ginger, cinnamon, baking soda, and salt in a large bowl. Add the molasses and soy milk and mix well. Add 1 cup of flour and mix well. Mix in enough of the remaining flour to make a very stiff dough (if mixing by hand, knead the dough to thoroughly mix the flour).

Preheat oven to 275°F. Lightly mist two or three baking sheets with vegetable oil spray, then dust them with flour. On a floured surface, roll a portion of the dough with a flour-dusted rolling pin to a 1⁄16-inch thickness. Cut the dough into shapes with flour-dusted cookie cutters or a flour-dusted knife. Using a metal spatula, carefully transfer the cookies to the baking sheets. Bake until the edges are dry and the centers give just slightly when pressed, about 20 minutes. Allow to cool on a baking sheet for 5 minutes, then transfer with a spatula to a wire rack to cool. Once cooled, store in an airtight container.

Per cookie: 35 calories; 1 g protein; 8 g carbohydrate; 0 g fat; 42 mg sodium

Pumpkin Spice Cookies

Makes thirty-six 3-inch cookies

These plump, moist cookies are easy to make and delicious.

3 cups whole wheat pastry flour
4 teaspoons sodium-free baking powder
1 teaspoon baking soda
1 teaspoon salt
2 teaspoons cinnamon
½ teaspoon nutmeg
1 15-ounce can solid-pack pumpkin (about 2 cups)
¾ cup sugar
½ cup molasses
1 cup soy milk, rice milk, or water
1 cup raisins

Preheat oven to 350°F. Mix the flour, baking powder, baking soda, salt, and spices together. In a separate bowl combine the pumpkin, sugar, molasses, and milk or water. Combine the two mixtures, then stir in the raisins. Drop by tablespoonfuls onto an oil-sprayed baking sheet. Bake 15 minutes, or until lightly browned. Remove from baking sheet with a spatula and place on a rack to cool. Once cooled, store in an airtight container in the refrigerator.

Per cookie: 80 calories; 2 g protein; 18 g carbohydrate; 0 g fat; 110 mg sodium

✳ Banana Bundt Cake ✳

Makes one Bundt cake or one 9 × 13-inch sheet cake

This cake is delicious plain or frosted with Date Butter Frosting (recipe follows).

> 3 cups whole wheat pastry flour
> 2 teaspoons baking soda
> ¾ teaspoon salt
> 1 cup wheat germ
> 6 ripe bananas, mashed (about 2½ cups)
> ¾ cup sugar
> 1 cup soy milk
> 2 teaspoons vanilla
> ½ cup finely chopped dates or date pieces

Preheat oven to 350°F. Mix the flour, baking soda, salt, and wheat germ in a bowl. In another bowl, mash the bananas and mix in the sugar, milk, and vanilla. Combine the flour mixture with the banana mixture. Add the dates and stir to mix. Spread into an oil-sprayed Bundt pan (or 9 × 13-inch baking dish) and bake for 55 minutes, until a toothpick inserted into the center comes out clean.

Per 1-inch slice: 207 calories; 6 g protein; 42 g carbohydrate; 1 g fat; 210 mg sodium

The Recipes

❋ Date Butter Frosting ❋

Makes about 3 cups

This frosting remains soft and spreadable.

2 cups soy milk
1 cup chopped pitted dates or date pieces
2 tablespoons cornstarch
1 teaspoon vanilla
½ teaspoon coconut extract*
¼ teaspoon salt
½ cup Corn Butter (page 200)

Process the soy milk, dates, and cornstarch in a blender until smooth, 2 to 3 minutes on high. Transfer to a saucepan and heat, stirring constantly, until the mixture bubbles and thickens (it should be the consistency of pudding). Remove from heat and stir in the vanilla, coconut extract, and salt. When cool, mix in the Corn Butter.

Per ½ cup: 134 calories; 3 g protein; 25 g carbohydrate; 2.5 g fat; 156 mg sodium

❋ Cranberry Corn Bread ❋

Makes 2 loaves

This bread is perfect for the holidays, when fresh cranberries are available. You can make it at other times of the year by substituting dried cranberries that have been soaked in warm water until soft.

---

*Frontier Naturals has a line of natural flavorings that is sold in natural food stores.

1 6-ounce can frozen orange juice concentrate,
   thawed
1 tablespoon lemon juice
2 cups whole wheat pastry flour
1 cup cornmeal
2 teaspoons baking soda
½ teaspoon salt
¾ cup corn syrup, rice syrup, or similar liquid
   sweetener
½ cup chopped walnuts (optional)
3 cups fresh cranberries (one 12-ounce bag)

Preheat oven to 350° F. Pour the orange juice concentrate into a measuring cup that holds 2 cups or more. Add the lemon juice and enough water to make 1½ cups. In a large bowl stir the flour, cornmeal, baking soda, and salt together. Add the orange juice and corn syrup. Stir to mix, then stir in the walnuts and cranberries. Do not overmix. Spoon into two nonstick or oil-sprayed loaf pans and bake for 1 hour. Let stand 5 minutes, then remove from pan and cool on a rack.

Per ½-inch slice with nuts: 95 calories; 2 g protein; 19 g carbohydrate; 1 g fat; 42 mg sodium
Per ½-inch slice without nuts: 83 calories; 2 g protein; 19 g carbohydrate; 0 g fat; 42 mg sodium

Product Guide/ Resources

S UPERMARKETS AND natural food stores offer a wide array of useful ingredients and ready-to-eat foods. Following is a partial list of products to help you get started. Some of these foods are high in sodium, so be sure to check the label if you are on a sodium-restricted diet. The manufacturer's contact information has been included to help you locate products.

BREAKFAST CEREALS

Wheatena	American Home Foods New York, NY 10017
Puffed Wheat, Puffed Corn, Puffed Millet, Nature O's, Multigrain Flakes, Corn Flakes, Amaranth Flakes, Kamut Flakes	Arrowhead Mills Box 2059 Hereford, TX 79045
Barbara's Shredded Wheat, Multigrain Shredded Spoonfuls	Barbara's Bakery Inc. Petaluma, CA 94954
Zoom	Continental Mills, Inc. Kent, WA 98032
Cheerios, Kix, Total Corn Flakes, Whole Grain Total, Total Raisin Bran	General Mills, Inc. Minneapolis, MN 55440
Health Valley cereals: Fat-Free Granola, Health Valley Organic Fiber 7 Flakes, Oat Bran Flakes, Amaranth Flakes	Health Valley Foods 16100 Foothill Blvd. Irwindale, CA 91706

Kashi, Puffed Kashi

Kashi Co.
La Jolla, CA 92038

Kellogg's Corn Flakes, Crispix,
Rice Krispies, Special K

Kellogg USA
Battle Creek, MI 49016

Nabisco Shredded Wheat,
Shredded Wheat 'N Bran, Grape-Nuts,
100% Bran, Cream of Rice

Kraft General Foods
White Plains, NY 10625

Malt-O-Meal

Malt-O-Meal Company
Minneapolis, MN 55402

Nature's Path cereals:
Millet Rice, Multigrain, Corn Flakes,
Rice Bran Flakes, Heritage Cereal

Nature's Path Foods
Delta, BC, Canada V4G 1E8

Oatios, Multi-Bran Flakes, Corn Flakes

New Morning
Acton, MA 01720

McCann's Quick Cooking Irish Oatmeal

Odlum Salling
Kildare, Ireland

Pacific Grain Nutty Rice Cereal

Pacific Grain Products
P.O. Box 2060
Woodland, CA 95776

Old Fashioned Quaker Oats, Quick Oats,
Multi Grain Oatmeal, Oat Bran Cereal

Quaker Oats Co.
Chicago, IL 60604

Wheat Chex, Corn Chex, Rice Chex

Ralston Foods
P.O. Box 618
St. Louis, MO 63188

Erewhon Brown Rice Cream Cereal

U.S. Mills
Omaha, NB 68111

Weetabix, Grainfields Cereals,	Weetabix Co.
Wheat Flakes, Cornflakes,	20 Cameron Street
Raisin Bran, Brown Rice	Clinton, MA 01510

FLATBREADS AND TORTILLAS

Cedarlane Fat Free Whole Wheat Tortillas,	Cedarlane Foods
Cedarlane Whole Wheat Pita Bread	1804 East 22nd St.
	Los Angeles, CA 90058

| Whole Wheat Bible Bread | Garden of Eatin' |
| | Los Angeles, CA 90029 |

Mission Fat Free Whole Wheat Tortillas	Mission Foods
	1159 Cottonwood Ln.
	Irving, TX 75038

RICE AND OTHER GRAINS

| Kasha (buckwheat groats) | Burnett Mills |
| | Penn Yan, NY 14527 |

Fantastic Foods Brown Rice Pilaf,	Fantastic Foods
Arborio Rice, Basmati Rice, Couscous,	1250 N. McDowell Blvd.
Whole Wheat Couscous, Basmati Rice,	Petaluma, CA 94954
Tabouli	

Nile Spice Couscous,	Nile Spice
Nile Spice Rozdali	Box 20581
	Seattle, WA 98102

| Texmati Brown Rice | RiceTec |
| | Alvin, TX 77511 |

Casbah Basmati Rice	Sahara Natural Foods
	2820 8th St.
	Berkeley, CA 94710

PASTA

Ronzoni pasta products

A. Foods Corp.
Hershey, PA 17033

De Boles Natural Pasta

Deboles Nutritional Foods
Garden City Park, NY 11040

DeCecco Penne Pasta

DeCecco
Fara S. Martino (CH)
Italy

U.S. Importer of DeCecco:
Prodotti Mediterraner, Inc.
712 Fifth Ave.
New York, NY 10019

Golden Grain Pasta (some types
contain egg; check labels)

Golden Grain Co.
San Leandro, CA 94578

Michelle's natural pastas (wide
variety of flavors,
all eggless)

Mrs. Leeper's
San Diego, CA 92127

PASTA SAUCE

Healthy Choice pasta sauces

ConAgra
Omaha, NE 68102

Millina's Finest pasta sauces
(several varieties, all with
organic ingredients, all fat-free)

Organic Food Products
P.O. Box 1510
Freedom, CA 95019

Weight Watchers Spaghetti Sauce

Weight Watchers Food Co.
P.O. Box 57
Pittsburgh, PA 15230

BEANS

Dennison's Chili Beans

American Home Foods
Vacaville, CA 95688

Bush's Vegetarian Beans

Bush Brothers & Co.
P.O. Box 52330
Knoxville, TN 37950

Fantastic Foods Rice & Beans in a Cup:
Caribbean, Tex Mex, Cajun, etc.;
Fantastic Foods Black Bean Flakes,
Pinto Bean Flakes, Hummus (some
flavors contain oil or butter; check label)

Fantastic Foods
1250 N. McDowell Blvd.
Petaluma, CA 94954

Health Valley Spicy Vegetarian Chili

Health Valley Foods
16100 Foothill Blvd.
Irwindale, CA 91706

Heinz Vegetarian Beans

H.J. Heinz Co.
Pittsburgh, PA 15212

Rosarita No Fat Refried Beans

Hunt-Wesson Foods
P.O. Box 4800
Fullerton, CA 92634

Bearitos Fat Free Beans: Refried Beans,
Refried Black Beans, Baked Beans,
Beans & Rice, etc.; Vegetarian Chili

Little Bear Organic Foods
Carson, CA 90746

Las Palmas No Fat Refried Beans

Ramirez Feraud Chili Co.
P.O. Box 54841
Los Angeles, CA 90054

S&W Maple Sugar Beans,
Pinquitos, Garbanzo Beans,
Deli Classics Marinated Bean Salad

S&W Fine Foods
San Ramon, CA 94583

Taste Adventure Dried Bean Flakes,
Black Beans, Pinto Beans

Will-Pak Foods
San Pedro, CA 90731

SOUPS

Bernard Jensen's Broth or Seasoning,
Quik Sip Boullion Concentrate

Bernard Jensen's Products
Solano Beach, CA 92075

Swanson Clear Vegetable Broth

Campbell Soup Co.
Camden, NJ 08103

Dr. McDougall's Soups:
Minestrone, Split Pea, etc.

Dr. McDougall's Right Foods
101 Utah Ave.
S. San Francisco, CA 94080

Shari's Bistro Organic Gourmet Soups:
Spicy French Green Lentil,
Tomato with Roasted Garlic,
Great Plains Split Pea, etc.

Fair Exchange, Inc.
P.O. Box 534
Dexter, MI 48130

Fantastic Foods Hearty Soup Cups,
Fantastic Foods Couscous Cups,
Fantastic Ramen Noodles Soup Cups

Fantastic Foods
1250 N. McDowell Blvd.
Petaluma, CA 94954

Health Valley fat-free soups,
Health Valley Fat Free Soup in a Cup
(wide selection of flavors)

Health Valley Foods
16100 Foothill Blvd.
Irwindale, CA 91706

Bearitos Homestyle fat-free soups:
Hearty Tex Mex, Lentil, Minestrone,
Southwest Vegetable, etc.

Little Bear Organic Foods
Carson, CA 90746

Vecon Vegetable Boullion

Norganic Foods
Norwalk, CA 90650

Chef's Classics Instant Soups

Pacific Foods of Oregon
19480 SW 97th Ave.
Tualatin, OR 97062

Soken Ramen:
Buckwheat, Brown Rice, Spicy
Dragon, etc.

Sokensha Co.
P.O. Box 883033
San Francisco, CA 94188

The Spice Hunter Soup Cups: Szechuan
Noodle, Hunan Noodle, French Country
Lentil, etc. (lower in sodium than most
commercial soups)

The Spice Hunter
San Luis Obispo, CA 93401

VegeX Vegetable Boullion

VegeX Co.
Flemington, NJ 08822

Westbrae Ramen

Westbrae Natural Foods
Commerce, CA 90040

Taste Adventure Dried Soups: Black
Bean, Split Pea, Navy Bean, etc.

Will-Pak Foods
San Pedro, CA

MEATLIKE PRODUCTS

Vegan Original Boca Burger

Boca Burger Co.
Fort Lauderdale, FL 33305

Vegetable Jerky

Garden of Eatin'
Los Angeles, CA 90029

Harvest Burger (mail order)

Harvest Direct
Knoxville, TN 37901-0988

Savory Seitan: Barbecue Flavor, Teriyaki
Flavor; Smart Dogs, Light Burgers, Vegi
Bologna, Gimme Lean (Beef Style,
Sausage Style), Smart Deli Thin Slices

Lightlife Foods
P.O. Box 870
Greenfield, MA 01302

Heartline Lite Textured Vegetable
Protein: Beef, Chicken, Canadian Bacon,
Pepperoni

Lumen Foods
Lake Charles, LA 70601

Green Giant Harvest Burger,	Pillsbury Co.
Green Giant Harvest Burger for Recipes	2866 Pillsbury Center
	Minneapolis, MN 55402
Meat Free Tender Deli Cuts	Soyco/Division of Galaxy Foods
	2441 Viscount Flow
	Orlando, FL 32809
Superburgers	Turtle Island Foods
	P.O. Box 176
	Hood River, OR 97031
Prime Burger, White Wave Baked Tofu,	White Wave Inc.
White Wave Seitan, Meatless Healthy	6123 Arapahoe
Franks, Meatless Jumbo Franks,	Boulder, CO 80303
Meatless Sandwich Slices,	
Chick'n Burger	
GardenVegan,	Wholesome and Hearty
Gardenburger Hamburger Style	Foods, Inc.
	2422 SE Hawthorne Blvd.
	Portland, OR 97214
Wild Dogs	Wildwood Natural Foods
	135 Bolinas Rd.
	Fairfax, CA 94930
Natural Touch Vegan Burger,	Worthington Foods, Inc.
All Vegetable Burger Crumbles,	900 Proprietors Rd.
Natural Touch Vegan Burger Crumbles,	Worthington, OH 43085
Natural Touch Vegan Sausage Crumbles	
Yves Veggie Cuisine Burger Burgers,	Yves Fine Foods
Yves Garden Vegetable Patties,	Vancouver, BC, Canada V6A
Veggie Wieners, Yves Chili Dogs,	2A8

Yves Tofu Dogs, Yves Veggie Pepperoni,
Yves Canadian Veggie Bacon,
Yves Deli Slices, Yves Breakfast Links,
Yves Just Like Ground!

NONDAIRY MILKS AND RELATED PRODUCTS

Natural food stores and supermarkets offer a wide selection of nondairy milk products that can be used in cooking, on cereal, and as beverages. These include soy milks, rice milks, oat milks, multigrain milks, and even potato milks! Their flavors vary widely, providing options for different tastes and uses. There are reduced-fat as well as enriched versions, refrigerated as well as nonrefrigerated varieties, and a growing number of powdered versions. In general, soy milks tend to be thick and creamy, with a distinctive soy flavor, while rice milks have a lighter consistency and flavor. Check the labels and choose varieties that have 3 grams of fat or less per serving.

EdenSoy®

Eden Foods, Inc.
800-248-0320

Ener-G Egg Replacer

Ener-G Foods
800-331-5222

Fat-Free Soy Moo®

Health Valley Foods
800-423-4846

Rice Dream®

Imagine Foods
415-327-1444

Better Than Milk?®,
Soymage Casein Free Parmesan

Sovex Foods, Inc.
800-227-2320

VitaSoy®

VitaSoy, Inc.
800-VITASOY

WestSoy®

Westbrae Natural Foods
800-SOY-MILK

Silk™   White Wave, Inc.
                                           303-443-3470

SALAD DRESSINGS, CONDIMENTS, AND MISCELLANEOUS INGREDIENTS

Cascadian Farm Spreadable Fruit   Cascadian Farm
                                           Rockport, WA 98283

Cook's Fat Free Dressings: Italian Garlic Cook's Classics
Gusto, Dijon, Country French Classics   P.O. Box 338
(be sure to select only fat-free varieties) Palo Alto, CA 94301

Emes Jel (nonanimal gelling agent)   Emes Kosher Products
                                           P.O. Box 833
                                           Lombard, IL 60148

Featherweight Baking Powder   Estee Corp.
(sodium-free baking powder)   Parsippany, NJ 07054-1094

Good Seasons Fat-Free Dressing Mix   General Foods Corp.
                                           White Plains, NY 10625

Weight Watcher's Italian Salad Dressing   H.J. Heinz Co.
                                           Pittsburgh, PA 15212

Kitchen Bouquet Seasoning Sauce   HVP Co.
                                           Oakland, CA

Kikkoman Milder Soy Sauce   Kikkoman
(less sodium than regular or lite)   Walworth, WI 53184

La Victoria Salsa   La Victoria Foods
                                           City of Industry, CA 91744

Bragg's Liquid Aminos (similar to   Live Food Products
soy sauce, with less sodium)   Box 7
                                           Santa Barbara, CA 93102

Seasoned Rice Vinegar	Marukan Vinegar Inc. Paramount, CA 90723
Seasoned Rice Vinegar	Nakano San Francisco, CA 94124
Fat-free Nayonaise	Nasoya Foods 1 New England Way Ayer, MA 01432
New Morning Apple Butter	New Morning Acton, MA 01720
Old El Paso Salsa	Old El Paso Foods Anthony, TX 88021
Pace Picante Sauce	Pace Foods P.O. Box 12636 San Antonio, TX 78212
Polaner Spreadable Fruit	Polaner Inc. Roseland, NJ 07068
Pritikin Salad Dressing	Pritikin Systems Chicago, IL 60604
Sorrell Ridge Spreadable Fruit	Sorrell Ridge Farm Port Reading, NJ 07064
S&W Vintage Lites Oil-Free Dressings: Raspberry Blush, Red Wine & Herb	S&W Foods San Ramon, CA 94583
Paula's No-Fat Dressings: Roasted Garlic, Garden Tomato, Toasted Onion, etc.	Sweet Adelaide Enterprises Hawthorne, CA 90250

CRACKERS AND SNACKS

VeraCruz Baked Tortilla Rounds

Alex Foods
Anaheim, CA 92803

Louise's Fat Free Potato Chips

ATGTBT Inc.
Louisville, KY 40299

Multigrain Crackers

Auburn Farms
Sacramento, CA 95834

Baja Black Bean Dip

Garden of Eatin'
Los Angeles, CA 90029

Laura Scudder's Pretzels

Granny Goose Foods
Oakland, CA 94603

Guiltless Gourmet
Fat Free Tortilla Chips,
Bean Dips

Guiltless Gourmet
3709 Promontory Pt. Dr.
#131
Austin, TX 78744

The Fat Free Gourmet Tortilla Chips

Harry's Snacks
P.O. Box 815
Syosset, NY 11791

Health Valley Fat Free Crackers

Health Valley Foods
16100 Foothill Blvd.
Irwindale, CA 91706

Kavli Crispbread

Kavli
Bergen, Norway

Kettle Creek Baked Tortilla Chips

Kettle Creek Outfitters
Perham, MN 56573

Wasa Crispbread: Golden Rye, Light Rye, Hearty Rye (other flavors contain fat)	Liberty Richter Saddle Brook, NJ 07662
Bearitos Bean Dip	Little Bear Organic Foods Carson, CA 90746
Quaker Rice Cakes (some flavors contain cheese; check label)	Quaker Oats P.O. Box 49003 Chicago, IL 60604
Finn Crisp Crispbread	Shaffer Clarke & Co. Darien, CT 06820
Snyder's of Hanover Hard Pretzels (some flavors contain fat; check label)	Snyder's of Hanover Hanover, PA 17331
Stella D'Oro Fat Free Bread Sticks	Stella D'Oro Biscuit Co. Bronx, NY 10463
Tree of Life Fat Free Crackers: Garden Vegetable, Cracked Pepper, Toasted Onion, Garlic and Herb	Tree of Life St. Augustine, FL 32085
Venus Fat Free Crackers: Cracked Pepper, Garlic and Herb	Venus Wafers Hingham, MA 02043

Mail-Order Sources
for Selected Products

Bob's Red Mill Natural Foods, Inc.
5209 S.E. International Way
Milwaukie, OR 97222
telephone: 1 (800) 553-2258
Web site: www.bobsredmill.com
Whole grain flours, cereals, and mixes

Dixie Diners' Club
telephone: 1 (800) 233-3668
fax: 1 (713) 688-4993
Web site: http://www.dixieusa.com
email: info@dixieusa.com
Wide assortment of fat-free vegan products

Dr. McDougall's Right Foods
101 Utah Avenue
S. San Francisco, CA 94080
telephone: 1 (415) 635-6000
Instant breakfasts, soups, bean cups, and meals

Harvest Direct
telephone: 1 (800) 835-2867
fax: 1 (423) 523-3372

email: harvest@slip.net
Vegan burger mixes and numerous textured vegetable
protein products

The Mail Order Catalog
telephone: 1 (800) 695-2241
fax: 1 (931) 964-3518
Web site: http://222.healthy-eating.com
email: catalog@usit.net
Textured vegetable protein products, seitan mixes,
nutritional yeast, vegetarian broth mixes, and cook-
books

Will-pak Foods
telephone: 1 (800) 874-0883
Web site: www.tasteadventure.com
Taste Adventure soups, dried bean flakes, and instant
meals

References

CHAPTER ONE: THE GENE SEARCH PAYS OFF

1. Bouchard C, Pérusse L, Rice T, and Rao DC, "The genetics of human obesity," in Bray GA, Bouchard C, and James WPT, *Handbook of Obesity* (New York: Marcel Dekker, Inc., 1998).

2. Stunkard AJ, Harris JR, Pedersen NL, and McClearn GE, "The body-mass index of twins who have been reared apart," *N Engl J Med* 322 (1990): 1483–87.

3. Comuzzie AG and Allison DB, "The search for human obesity genes," *Science* 280 (1998): 1374–77.

4. Collin GB, Marshall JD, Cardon LR, and Nishina PM, "Homozygosity mapping at Alstrom syndrome to chromosome 2p," *Hum Mol Genet* 6 (1997): 213–19.

5. Chagnon YC, Pérusse L, and Bouchard C, "The human obesity gene map: The 1997 update," *Obes Res* 6 (1998): 76–92.

6. Montague CT, Farooqi IS, Whitehead JP et al., "Congenital leptin deficiency is associated with severe early-onset obesity in humans," *Nature* 387 (1997): 903–8.

7. Stunkard AJ, Sorensen TIA, Hanis C et al., "An adoption study of human obesity," *N Engl J Med* 314 (1986): 193–98.

8. Fabsitz RR, Carmelli D, and Hewitt JK, "Evidence of independent genetic influences on obesity in middle age," *Int J Obes* 16 (1992): 657–66.

9. Rice T, Borecki IB, Bouchard C, and Rao DC, "Segregation analysis of fat mass and other body composition measures derived from underwater weighing," *Am J Hum Genet* 52 (1993): 967–73.

10. Rice T, Tremblay A, Dériaz O, Pérusse L, Rao DC, and Bouchard C, "A major gene for resting metabolic rate unassociated with body composition: Results from the Quebec Family Study," *Obes Res* 4 (1996): 441–49.

11. Hasstedt SJ, Ramirez ME, Kuida H, and Williams RR, "Recessive inheritance of a relative fat pattern," *Am J Hum Genet* 45 (1989): 917–25.

12. Ness R, Laskarzewski P, and Price RA, "Inheritance of extreme overweight in black families," *Hum Biol* 63 (1991): 39–52.

13. Norman RA, Thompson DB, Foroud T et al., "Genomewide search for genes influencing percent body fat in Pima Indians: Suggestive linkage at chromosome 11q21-q22," *Am J Hum Genet* 60 (1997): 166–73.

14. Pérusse L and Bouchard C, "Genetics of energy intake and food preferences," in Bouchard C, ed., *The Genetics of Obesity* (Boca Raton: CRC Press, 1994), 125–34.

15. Drewnowski A, "The genetics of taste and smell," in Simopoulos AP and Childs B, eds., *Genetic Variation and Nutrition* (Basel: S. Karger, 1989), 194.

16. Keim NL, Stern JS, and Havel PJ, "Relation between circulating leptin concentrations and appetite during a prolonged, moderate energy deficit in women," *Am J Clin Nutr* 68 (1998): 794–801.

17. Ravussin E, Pratley RE, Maffei M et al., "Relatively low plasma leptin concentrations precede weight gain in Pima Indians," *Nat Med* 3 (1997): 238–40.

18. Gura T, "Uncoupling proteins provide new clue to obesity's causes," *Science* 280 (1998): 1369–70.

19. Bouchard C, Dériaz O, Pérusse L, and Tremblay A, "Genetics of energy expenditure in humans," in Bouchard, *Genetics of Obesity*, 135–45.

20. Ravussin E, Lillioja S, Knowler WC et al., "Reduced rate of energy expenditure as a risk factor for body-weight gain," *N Engl J Med* 318 (1988): 467–72.

CHAPTER TWO: TASTE GENES: BROCCOLI AND CHOCOLATE

1. Drewnowski A, "Taste preferences and food intake," *Annu Rev Nutr* 17 (1997): 237–53.

2. Drewnowski A, Henderson SA, and Barratt-Fornell A, "Genetic sensitivity to 6-n-propylthiouracil and sensory responses to sugar and fat mixtures," *Physiol Behav* 63 (1998): 771–77.

3. Tepper BJ and Nurse RJ, "Fat perception is related to PROP taster status," *Physiol Behav* 61 (1997): 949–54.

4. Drewnowski A and Rock CL, "The influence of genetic taste markers on food acceptance," *Am J Clin Nutr* 62 (1995): 506–11.

5. Akela GD, Henderson SA, and Drewnowski A, "Sensory acceptance of Japanese green tea and soy products is linked to genetic sensitivity to 6-n-propylthiouracil," *Nutr Cancer* 29 (1997): 146–51.

6. Tepper BJ and Nurse RJ, "PROP taster status is related to the perception and preference for fat," *AChemS XIX Abstracts* (1997).

7. Anliker JA, Bartoshuk L, Ferris AM, and Hooks LD, "Children's food preferences and genetic sensitivity to the bitter taste of 6-n-propylthiouracil (PROP)," *Am J Clin Nutr* 54 (1991): 316–20.

8. Looy H and Weingarten HP, "Facial expressions and genetic sensitivity to 6-n-propylthiouracil predict hedonic response to sweet," *Physiol Behav* 52 (1992): 75–82.

9. Fischer R, Griffin F, and Rockey MA, "Gustatory chemoreception in man: Multidisciplinary aspects and perspectives," *Persp Biol Med* 9 (1966): 549–77.

10. Lucchina LA, Bartoshuk LM, Duffy VB, Marks LE, and Ferris AM, "6-n-propylthiouracil perception affects nutritional status of independent-living older females," *Chem Senses* 20 (1995): 735.

11. Drewnowski A, Henderson SA, and Shore AB, "Taste responses to naringin, a flavonoid, and the acceptance of grapefruit juice are related to genetic sensitivity to 6-n-propylthiouracil," *Am J Clin Nutr* 66 (1997): 391–97.

12. Fisher JO and Birch JL, "Fat preferences and fat consumption of 3-to-5-year-old children are related to parent adiposity," *J Am Diet Assoc* 95 (1995): 759–64.

13. Hill AJ and Heaton-Brown L, "The experience of food craving: A prospective investigation in healthy women," *J Psychosom Res* 38 (1994): 801–14.

14. Gendall KA, Joyce PR, and Sullivan PF, "Impact of definition of prevalence of food cravings in a random sample of young women," *Appetite* 28 (1997): 63–72.

15. Pelchat ML, "Food cravings in young and elderly adults," *Appetite* 28 (1997): 103–13.

16. Rozin P, Levine E, and Stoess C. "Chocolate craving and liking," *Appetite* 17 (1991): 199–212.

17. Morley JE and Levine AS, "The role of the endogenous opiates as regulators of appetite," *Am J Clin Nutr* 35 (1982): 757–61.

18. Drewnowski A, Krahn DD, Demitrack MA, Nairn K, and Gosnell BA, "Taste responses and preferences for sweet high-fat foods: Evidence for opioid involvement," *Physiol Behav* 51 (1992): 371–79.

19. Rozin P, "Family resemblance in food and other domains: The family paradox and the role of parental congruence," *Appetite* 16 (1991): 93–102.

20. Michell GF, Mebane AH, and Billings CK, "Effect of bupropion on chocolate craving," *Am J Psychiatry* 146 (1989): 119–20.

21. Mela DJ and Sacchetti DA, "Sensory preferences for fats: Relationships with diet and body composition," *Am J Clin Nutr* 53 (1991): 908–15.

22. Drewnowski A, "Why do we like fat?" *J Am Diet Assoc* 97 (suppl.) (1997): S58–62.

23. Nordin BEC, Need AG, Morris HA, and Horowitz M, "The nature and significance of the relationship between urinary sodium and urinary calcium in women," *J Nutr* 123 (1993): 1615–22.

24. Barnard ND, Scialli AR, Hurlock D, and Bertron P, "Diet and sex-hormone binding globulin, dysmenorrhea, and premenstrual symptoms," *Obstet Gynecol* 95 (2000): 245–50.

CHAPTER THREE: APPETITE AND THE LEPTIN GENE

1. Heymsfield SB, Greenberg AS, Fujioka K et al., "Recombinant leptin for weight loss in obese and lean adults: A randomized, controlled, dose-escalation trial," *JAMA* 282 (1999): 1568–75.

2. Rosenbaum M and Leibel RL, "The role of leptin in human physiology," *N Engl J Med* 341 (1999): 913–14.

3. Keim NL, Stern JS, and Havel PJ, "Relation between circulating leptin concentrations and appetite during a prolonged, moderate energy deficit in women," *Am J Clin Nutr* 68 (1998): 794–801.

4. Girard J, "Is leptin the link between obesity and insulin resistance?" *Diab Metab* 23 (1997): 16–24.

5. Cella F, Adami GF, Giordano G, and Cordera R, "Effects of dietary restriction on serum leptin concentration in obese women," *Int J Obes* 23 (1999): 494–97.

6. Wadden TA, Considine RV, Foster GD, Anderson DA, Sarwer DB, and Caro JS, "Short- and long-term changes in serum leptin in dieting obese women: Effects of caloric restriction and weight loss," *J Clin Endocrin Metab* 83 (1998): 214–18.

7. Auwerx J and Staels B, "Leptin," *Lancet* 351 (1998): 737–42.

8. Farooqi IS, Jebb SA, Langmack G et al., "Effects of recombinant leptin therapy in a child with congenital leptin deficiency," *N Engl J Med* 341 (1999): 879–84.

9. Comuzzie AG, Hixson JE, Almasy L et al., "A major quantitative trait locus determining serum leptin levels and fat mass is located on human chromosome 2," *Nature Genet* 15 (1997): 273–76.

References · 327

10. Mehler PS, Eckel RH, and Donahoo WT, "Leptin levels in restricting and purging anorectics," *Int J Eat Disord* 26 (1999): 189–94.

11. Herman CP and Mack D, "Restrained and unrestrained eating," *J Personality* 43 (1975): 647–60.

12. Ludwig DS, Pereira MA, Kroenke CH et al., "Dietary fiber, weight gain, and cardiovascular disease risk factors in young adults," *JAMA* 282 (1999): 1539–46.

13. Weaver CM and Plawecki KL, "Dietary calcium: Adequacy of a vegetarian diet," *Am J Clin Nutr* 59 (suppl.) (1994): 1238S–41S.

CHAPTER FOUR: THE FAT-BUILDING GENE

1. Rosenbaum M and Leibel RL, "The role of leptin in human physiology," *N Engl J Med* 341 (1999): 913–14.

2. Ristow M, Muller-Wieland D, Pfeiffer A, Krone W, and Kahn CR, "Obesity associated with a mutation in a genetic regulator of adipocyte differentiation," *N Engl J Med* 339 (1998): 953–59.

3. Robertson SM, Cullen KW, Baranowski J, Baranowski T, Hu S, and de Moor C, "Factors related to adiposity among children aged 3 to 7 years," *J Am Diet Asso* 99 (1999): 938–43.

4. Dayton S, Hashimoto S, Dixon W, and Pearce ML, "Composition of lipids in human serum and adipose tissue during prolonged feeding of a diet high in unsaturated fat," *J Lipid Res* 76 (1966): 103–11.

5. Field CJ and Clandinin MT, "Modulation of adipose tissue fat composition by diet: A review," *Nutr Research* 4 (1984): 743–55.

6. Linscheer WG and Vergroesen AJ, "Lipids," in Shils ME, Olson JA, and Shike M, eds., *Modern Nutrition in Health and Disease* (Philadelphia: Lea & Febiger, 1994), 64.

7. Holman RT, Johnson SB, and Hatch TF, "A case of human linolenic acid deficiency involving neurological abnormalities," *Am J Clin Nutr* 35 (1982): 617–23.

8. Barnard ND, Scialli AR, Hurlock D, and Bertron P, "Diet and sex-hormone binding globulin, dysmenorrhea, and premenstrual symptoms," *Obstet Gynecol* 95 (2000): 245–50.

9. McDowell MA, Briefel RR, Alaimo K et al., "Energy and macronutrient intakes of persons ages 2 months and over in the United States: Third national health and nutrition examination survey, phase 1, 1988–91, *Advance Data* (National Center for Health Statistics, Centers for Disease Control and Prevention) 255 (October 24, 1994).

10. Cees de Graaf J, Drijvers JMM, and Zimmermanns NJH, "Energy and fat compensation during long-term consumption of reduced-fat products," *Appetite* 29 (1997): 305–23.

11. Acheson KJ, Flatt JP, and Jéquier E, "Glycogen synthesis versus lipogenesis after a 500-gram carbohydrate meal in man," *Metabolism* 31 (1982): 1234–40.

12. Acheson KJ, Schutz Y, Bessard T, Ravussin E, Jéquier E, and Flatt JP, "Nutritional influences on lipogenesis and thermogenesis after a carbohydrate meal," *Am J Physiol* 246 (1984): E62–E70.

13. Swinburn B and Ravussin E, "Energy balance or fat balance?" *Am J Clin Nutr* 57 (suppl) (1993): 766S–71S.

14. Horton TJ, Drougas H, Brachey A, Reed GW, Peters JC, and Hill JO, "Fat and carbohydrate overfeeding in humans: Different effects on energy storage," *Am J Clin Nutr* 62 (1995): 19–29.

15. Eckel RH, "Lipoprotein lipase: A multifunctional enzyme relevant to common metabolic diseases," *N Engl J Med* 320 (1989): 1060–68.

CHAPTER FIVE: FAT-BURNING: TURNING THE FLAME HIGHER

1. McArdle WD, Katch FI, and Katch VL, *Exercise Physiology: Energy, Nutrition, and Human Performance,* 4th ed. (Baltimore: Williams & Wilkins, 1996), 151.

2. Rice T, Tremblay A, Dériaz O, Pérusse L, Rao DC, and Bouchard C, "A major gene for resting metabolic rate unassociated with body composition: Results from the Quebec Family Study," *Obes Res* 4 (1996): 441–49.

3. Ravussin E, Lillioja S, Knowler WC et al., "Reduced rate of energy expenditure as a risk factor for body-weight gain," *N Engl J Med* 318 (1993): 467–72.

4. Fleury C, Neverova M, Collins S et al., "Uncoupling protein 2: A novel gene linked to obesity and hyperinsulinemia," *Nature Genet* 15 (1997): 269–72.

5. Bouchard C, Pérusse L, Chagnon YC, Warden C., and Ricquier D, "Linkage between markers in the vicinity of the uncoupling protein 2 gene and resting metabolic rate in humans," *Hum Mol Genet* 6 (1997): 1887–89.

6. Tappy L, Felber JP, and Jéquier E, "Energy and substrate metabolism in obesity and postobese state," *Diabetes Care* 14 (1991): 1180–88.

7. Jung RT, Shetty PS, and James WP, "The effect of refeeding after semistarvation on catecholamine and thyroid metabolism," *Int J Obes* 4 (1980): 95–100.

8. Wadden TA, Foster GD, Letizia KA, and Mullen JL, "Long-term effects of dieting on resting metabolic rate in obese outpatients," *JAMA* 264 (1990): 707–11.

9. Leibel RL, Rosenbaum M, and Hirsch J, "Changes in energy expenditure resulting from altered body weight," *N Engl J Med* 332 (1995): 621–28.

10. Henson LC, Poole DC, Donahoe CP, and Heber D, "Effects of exercise training on resting energy expenditure during caloric restriction," *Am J Clin Nutr* 46 (1987): 893–99.

11. Schutz Y, Flatt JP, and Jéquier E, "Failure of dietary fat intake to promote fat oxidation: A factor favoring the development of obesity," *Am J Clin Nutr* 50 (1989): 307–14.

12. Ravussin E, Acheson KJ, Vernet O, Danforth E, and Jéquier E, "Evidence that insulin resistance is responsible for the decreased thermic effect of glucose in human obesity," *J Clin Invest* 76 (1985): 1268–73.

13. Kern PA, "Potential role of TNFα and lipoprotein lipase as candidate genes for obesity," *J Nutr* 127 (1997): 1917S–22S.

14. Richelsen B, Pedersen SB, Moller-Pedersen T, Schmitz O, Moller N, and Borglum JD, "Lipoprotein lipase activity in muscle tissue influenced by fatness, fat distribution, and insulin in obese females," *Eur J Clin Invest* 23 (1993): 226–33.

15. Ludwig DS, Pereira MA, Kroenke CH et al., "Dietary fiber, weight gain, and cardiovascular disease risk factors in young adults," *JAMA* 282 (1999): 1539–46.

16. Tappy L and Jéquier E, "Fructose and dietary thermogenesis," *Am J Clin Nutr* 58 (suppl.) (1993): 766S–70S.

17. Poehlman ET, Tremblay A, Nadeau A, Dussault J, Thériault G, and Bouchard C, "Heredity and changes in hormones and metabolic rates with short-term training," *Am J Physiol* 250 (1986a): E711–17.

18. Acheson KJ, Schutz Y, Bessard T, Ravussin E, Jéquier E, and Flatt JP, "Nutritional influences on lipogenesis and thermogenesis after a carbohydrate meal," *Am J Physiol* 246 (1984): E62–E70.

19. Segal KR, Edaño A, Blando L, and Pi-Sunyer FX, "Comparison of thermic effects of constant and relative calorie loads in lean and obese men," *Am J Clin Nutr* 51 (1990): 14–21.

20. Bouchard C, Dériaz O, Pérusse L, and Tremblay A, "Genetics of energy expenditure in humans," in Bouchard C, ed., *The Genetics of Obesity* (Boca Raton: CRC Press, 1994), 135–45.

21. Lovejoy JC, Windhauser MM, Rood JC, and de la Bretonne JA, "Effect of a controlled high-fat versus low-fat diet on insulin sensitivity and leptin levels in African-American and Caucasian women," *Metab* 47 (1998): 1520–24.

22. Poehlman ET, Arciero PJ, Melby CL, and Badylak SF, "Resting metabolic rate and postprandial thermogenesis in vegetarians and nonvegetarians," *Am J Clin Nutr* 48 (1988): 209–13.

23. Jenkins DJA, Ghafari H, Wolever TMS, Taylor RH, Jenkins AL, Barker HM et al., "Relationship between rate of digestion of foods and postprandial glycaemia," *Diabetologia* 22 (1982): 450–55.

24. Jenkins DJA, Josse RG, Jenkins AL, Wolever TMS, and Vuksan V, "Implications of altering the rate of carbohydrate absorption from the gastrointestinal tract," *Clin Invest Med* 18 (1995): 296–302.

25. Rasmussen O, Winther E, Arnfred J, and Hermansen K, "Comparison of blood glucose and insulin responses in non-insulin-dependent diabetic patients: Studies with spaghetti and potato taken alone and as part of a mixed meal," *Eur J Clin Nutr* 42 (1988): 953–61.

26. Jenkins DJA, Wolever TMS, and Jenkins AL, "Starchy foods and glycemic index," *Diabetes Care* 11 (1988): 149–59.

27. Ludwig DS, Majzoub JA, Al-Zahrani A, Dallal GE, Blanco I, and Roberts SB, "High glycemic index foods, overeating, and obesity," *Pediatrics* 103 (1999): 656.

28. Holt S, Brand J, Soveny C, and Hansky J, "Relationship of satiety to postprandial glycaemic, insulin, and cholecystokinin responses," *Appetite* 18 (1992): 129–41.

29. Leathwood P and Pollet P, "Effects of slow-release carbohydrates in the form of bean flakes on the evolution of hunger and satiety in man," *Appetite* 10 (1988): 1–11.

30. Granfeldt Y, Hagander B, and Bjorck I, "Metabolic responses to starch in oat and wheat products: The importance of food structure, incomplete gelatinization, or presence of viscous dietary fiber," *Eur J Clin Nutr* 49 (1995): 189–99.

31. Wolever TMS, Jenkins DJA, Ocana AM, Rao VA, and Collier GR, "Second-meal effect: Low-glycemic-index foods eaten at dinner improve subsequent breakfast glycemic response," *Am J Clin Nutr* 48 (1988): 1041–47.

32. Holt SHA, Brand Miller JC, and Petocz P, "An insulin index of foods: The insulin demand generated by 1000-kJ portions of common foods," *Am J Clin Nutr* 66 (1997): 1264–76.

33. Fabry P and Tepperman J, "Meal frequency—a possible factor in human pathology," *Am J Clin Nutr* 25 (1970): 1059–68.

CHAPTER SIX: HOW GENES INFLUENCE YOUR EXERCISE

1. Bassett DR, Jr., "Skeletal muscle characteristics: Relationships to cardiovascular risk factors," *Med Sci Sports Exerc* 26 (1994): 957–66.

2. Simsolo RB, Ong JM, and Kern PA, "The regulation of adipose tissue and muscle lipoprotein lipase in runners by detraining," *J Clin Invest* 92 (1993): 2124–30.

3. Kiens B, Lithell H, Mikines KJ, and Richter EA, "Effects of insulin and exercise on muscle lipoprotein lipase activity in man and its relation to insulin action," *J Clin Invest* 84 (1989): 1124–29.

4. Poehlman ET, Melby CL, and Badylak SF, "Resting metabolic rate and postprandial thermogenesis in highly trained and untrained males," *Am J Clin Nutr* 47 (1988): 793–98.

5. Bouchard C, Dériaz O, Pérusse L, and Tremblay A, "Genetics of energy expenditure in humans," in Bouchard C, ed., *The Genetics of Obesity* (Boca Raton: CRC Press, 1994), 135–45.

6. Bouchard C, Tremblay A, Després JP et al., "The response to exercise with constant energy intake in identical twins," *Obes Res* 2 (1994): 400–410.

7. Jakicic JM, Winters C, Lang W, and Wing RR, "Effects of intermittent exercise and use of home exercise equipment on adherence, weight loss, and fitness in overweight women," *JAMA* 282 (1999): 1554–60.

8. Miller WC, Koceja DM, and Hamilton EF, "A meta-analysis of the past 25 years of weight loss research using diet, exercise, or diet plus exercise intervention," *Int J Obesity* 21 (1997): 941–47.

9. Williams C, "Diet and sports performance," in Harries M, Williams C, Stanish WD, and Micheli LJ, eds., *Oxford Textbook of Sports Medicine* (Oxford: Oxford University Press, 1998), 80–92.

10. Kaplan RM, Patterson TL, Sallis JF, Jr., and Nader PR, "Exercise suppresses heritability estimates for obesity in Mexican-American families," *Addict Behav* 14 (1989): 581–88.

CHAPTER SEVEN: THE THREE-WEEK DIET MAKEOVER

1. Bertino M, Beauchamp GK, and Engelman K, "Increasing dietary salt alters salt taste preference," *Physiol Behav* 38 (1986): 203–13.

2. Mattes RD, "Fat preference and adherence to a reduced-fat diet," *Am J Clin Nutr* 57 (1993): 373–81.

3. Nicholson AS, Sklar M, Barnard ND, Gore S, Sullivan R, and Browning S, "Toward improved management of NIDDM: A randomized, controlled, pilot intervention using a low-fat, vegetarian diet," *Prev Med* 29 (1999): 87–91.

4. Barnard ND, Scialli AR, Hurlock D, and Bertron P, "Diet and sex-hormone-binding globulin, dysmenorrhea, and premenstrual symptoms," *Obstet Gynecol* 95 (2000): 245–50.

5. Barnard ND, Akhtar A, and Nicholson A, "Factors that facilitate compliance to lower fat intake," *Arch Fam Med* 4 (1995): 153–58.

6. Fischer R, Griffin F, and Rockey MA, "Gustatory chemoreception in man: Multidisciplinary aspects and perspectives," *Perspectives Biol Med* 9 (1966): 549–77.

CHAPTER EIGHT: FOOD CHOICES FOR OPTIMAL WEIGHT CONTROL

1. Liddle RA, Goldstein RB, and Saxton J, "Gallstone formation during weight-reduction dieting," *Arch Intern Med* 149 (1989): 1750–53.

2. Kamrath RO, Plummer LJ, Sadur CN et al., "Cholelithiasis in patients treated with a very low calorie diet," *Am J Clin Nutr* 56 (1992): 255S–57S.

3. Weinsier RL, Wilson LJ, and Lee J, "Medically safe rate of weight loss for the treatment of obesity: A guideline based on risk of gallstone formation," *Am J Med* 98 (1995): 115–17.

4. Syngal S, Coakley EH, Willett WC, Byers T, Williamson DF, and Colditz GA, "Long-term weight patterns and risk for cholecystectomy in women," *Ann Intern Med* 130 (1999): 471–77.

5. Ornish D, Brown SE, Scherwitz LW et al., "Can lifestyle changes reverse coronary heart disease?" *Lancet* 336 (1990): 129–33.

6. Hunninghake DB, Stein EA, Dujovne CA et al., "The efficacy of intensive dietary therapy alone or combined with lovastatin in outpatients with hypercholesterolemia," *New Engl J Med* 328 (1993): 1213–19.

7. Nicholson AS, Sklar M, Barnard ND, Gore S, Sullivan R, and Browning S, "Toward improved management of NIDDM: A randomized, controlled, pilot intervention using a low-fat, vegetarian diet," *Prev Med* 29 (1999): 87–91.

8. Smith CF, Burke LE, and Wing RR, "Young adults remain on vegetarian diet longer than on weight loss diets," *Ann Beh Med* 21 (suppl.) (1999): S090.

CHAPTER NINE: COMPLETE NUTRITION

1. Abelow BJ, Holford TR, and Insogna KL, "Cross-cultural association between dietary animal protein and hip fracture: A hypothesis," *Calif Tissue Int* 50 (1992): 14–18.

2. Remer T and Manz F, "Estimation of the renal net acid excretion by adults consuming diets containing variable amounts of protein," *Am Clin Nutr* 59 (1994): 1356–61.

3. Nordin BEC, Need AG, Morris HA, and Horowitz M, "The nature and significance of the relationship between urinary sodium and urinary calcium in women," *J Nutr* 123 (1993): 1615–22.

4. Massey LK and Whiting SJ, "Caffeine, urinary calcium, calcium metabolism, and bone," *J Nutr* 123 (1993): 1611–14.

5. Hopper JL and Seeman E, "The bone density of female twins discordant for tobacco use," *N Engl J Med* 330 (1994): 387–92.

6. Feskanich D, Willett WC, Stampfer MJ, and Colditz GA, "Milk, dietary calcium, and bone fractures in women: A 12-year prospective study," *Am J Publ Health* 87 (1997): 992–97.

CHAPTER TEN: ABOUT HIGH-PROTEIN DIETS

1. Holt SHA, Brand Miller JC, and Petocz P, "An insulin index of foods: The insulin demand generated by 1000-kJ portions of common foods," *Am J Clin Nutr* 66 (1997): 1264–76.

2. Remer T and Manz F, "Estimation of the renal net acid excretion by adults consuming diets containing variable amounts of protein," *Am Clin Nutr* 59 (1994): 1356–61.

3. World Cancer Research Fund/American Institute for Cancer Research, *Food, Nutrition, and the Prevention of Cancer: A Global Perspective* (Washington, D.C.: American Institute for Cancer Research, 1997), 499.

4. Ornish D, Brown SE, Scherwitz LW et al., "Can lifestyle changes reverse coronary heart disease?" *Lancet* 336 (1990): 129–33.

CHAPTER ELEVEN: CHILDREN AND THE FAT GENES

1. Ravelli ACJ, van der Meulen JHP, Osmond C, Barker DJP, and Bleker OP, "Obesity at the age of 50 in men and women exposed to famine prenatally," *Am J Clin Nutr* 70 (1999): 811–16.

2. Whitaker RC, Wright JA, Pepe MS, Seidel KD, and Dietz WH, "Predicting obesity in young adulthood from childhood and parental obesity," *N Engl J Med* 337 (1997): 869–73.

3. Robertson SM, Cullen KW, Baranowski J, Baranowski T, Hu S, and de Moor C, "Factors related to adiposity among children aged 3 to 7 years," *J Am Diet Asso* 99 (1999): 938–43.

4. Gazzaniga JM and Burns TL, "Relationship between diet composition and body fatness, with adjustment for resting energy expenditure and physical activity, in preadolescent children," *Am J Clin Nutr* 58 (1993): 21–28.

5. Oliveria SA, Ellison RC, Moore LL, Gillman MW, Garrahie EJ, and Singer MR, "Parent-child relationships in nutrient intake: The Framingham Children's Study," *Am J Clin Nutr* 56 (1992): 593–98.

6. Anderson JW, Johnstone BM, and Remley DT, "Breast-feeding and cognitive development: A meta-analysis," *Am J Clin Nutr* 70 (1999): 525–35.

7. Feskanich D, Willett WC, Stampfer MJ, and Colditz GA, "Milk, dietary calcium, and bone fractures in women: A 12-year prospective study," *Am J Publ Health* 87 (1997): 992–97.

8. Kauppila LI, "Can low-back pain be due to lumbar-artery disease?" *Lancet* 346 (1995): 888–89.

9. Barnard ND, Nicholson A, and Howard JL, "Medical costs attributable to meat consumption," *Prev Med* 24 (1995): 646–55.

10. Roberts SB, Savage J, Coward WA, Chew B, and Lucas A, "Energy expenditure and intake in infants born to lean and overweight mothers," *N Engl J Med* 318 (1988): 461–66.

11. Moore LL, Lombardi DA, White MJ, Campbell JL, Oliveria SA, and Ellison SA, "Influence of parents' physical activity levels on young children," *J Pediatr* 118 (1991): 215–19.

12. Robinson TN, "Reducing children's television viewing to prevent obesity," *JAMA* 282 (1999): 1561–67.

References

Index

Index

Fruits, 89
    after-meal burn and, 86
    fiber content, 50
    metabolism and, 82
    Muesli, 182
    protein content, 133
    Quick Breakfast Pudding, 182–183
    See also specific fruits

Gallstones, 118
Garbanzo beans
    Curried Potatoes and Chickpeas,
        268–269
    Garbanzo Salad Sandwich, 225
    Garbanzo Wraps, 232
    Red Pepper Hummus, 197
Garlic
    Garlic Bread I and II, 193–194
    Green Beans with Garlic, 260–261
Gazpacho, 244–245
Genes, 1
    for body shape, 16
    functioning of, 2, 15
    research on, 2
Genetics of weight, xii-xiii, 1–2
    abnormal conditions, 13–14
    complexity of, 17–18
    environment and, 12, 16–17
    ethnic groups, gene effects in, 16
    flexibility of genes, 2–3
    identification of fat genes and thin
        genes, 12–14
    research on, 10–12
    See also Appetite; Exercise; Insulin;
        Leptin; LPL; Metabolism; Taste
Genome scanning, 13
Giacometti, Alberto, 24
Ginger
    Gingerbread, 296–297
    Gingerbread Cookies, 303–304

Ginger Noodles, 271–272
Ginger Peachy Bread Pudding,
    301–302
Glycemic index, 91, 103
Glycogen, 67–68, 77, 79, 81, 103, 144
Grain dishes
    Always Great Brown Rice, 270
    Bulgur, 272
    Couscous, 273
    Ginger Noodles, 271–272
    Muesli, 182
    Multigrain Cereal, 179
    Multigrain Pancakes, 186
    Polenta, 273–274
    Quick Confetti Rice, 270–271
    Quinoa, 275
    Rolled Grain Cereal, 180
    Zucchini Corn Fritters, 274–275
    See also specific grains
Grain products
    after-meal burn and, 86–87
    fiber content, 51
    product guide/resources, 311
    protein content, 133
    Whole Grain Group, 119–120
Grapefruit, 29–30
Gravy, Brown, 201–202
Gravy, Mashed Potatoes and, 252

Heart disease, 49, 121
Herman, Peter, 44–45
Heterocyclic amines, 133–134, 145
High-protein diets, 130–131, 143–146
    dangers of, 145
    rationale for, 143–145
    research on, 146
Hoppin' John Salad, 220–221
Hostess cupcakes, 31
Hot sauce, 33
Human Genome Project, 2, 13

Hummus, Red Pepper, 197
Hunger, sensation of, 130

Ice cream, 31
Insulin, 4, 18, 20–21, 71–72
    after-meal burn and, 77–81, 82,
        83–84
    body fat and, 62, 80
    carbohydrates and, 71, 82
    exercise and, 83
    function of, 77
    genetics of insulin sensitivity, 84
    high-protein diets and, 144, 145
    LPL and, 62, 80
    malfunctioning insulin, problems of,
        79–81, 82, 85
    meal size and, 93
    protein and, 92
    tryptophan and, 35
Iron, 135–136

*Journal of the American Medical
    Association,* 47–48, 101

Kale
    Portuguese Kale Soup, 238
    Potatoes and Kale, 253
    Sesame Kale, 249
Kempner, Walter, 119–120
Ketones, 144, 145
Kidney disease, 131, 145

Lasagne Roll-ups, 291–292
Legume Group, 120
Legumes
    after-meal burn and, 87
    calcium content, 138, 139
    fiber content, 51
    protein content, 103, 132
    See also Beans

Lemon Curd, Fresh, 299–300
Lemon juice, 28
Lentils
    French Green Lentils, 263–264
    Red Lentil Curry, 262–263
Leptin, 1, 3, 18, 19–20
    abnormal blocking of, 14
    appetite, influence on, 42–44
    binge eating and, 44–47
    dieting and, 43–44
    diet makeover and, 118
    manufacture of, 15
    normal levels of, 44
    weight loss and, 41, 42–43
Lettuce
    Tofu, Lettuce, and Tomato Sandwich
        (TLT), 230
Lipoprotein lipase. *See* LPL
Liposuction, 42
Liver (food), 136
Low-calorie diets. *See* Dieting
LPL (lipoprotein lipase), 3, 18, 20
    childhood obesity and, 147, 149
    deficiency problem, 70
    diet makeover and, 119–121
    exercise's effect on, 97–98
    fat-storage function, 56–58, 62, 69–70
    insulin and, 62, 80
    in muscles, 57, 96

Magnesium, 34
Main dishes
    Black Bean Pueblo Pie, 279–280
    Eggplant Manicotti, 277–278
    Lasagne Roll-ups, 291–292
    Mexican Skillet Pie, 283–284
    No-Meat Loaf, 286–287
    Polenta Pizza, 276–277
    Portabellos with Collards and
        Cannelini Beans, 288–289

Nutrition (*cont.*)
carbohydrates and, 134
for children, 150–153
iron and, 135–136
protein and, 131–134
vegetarian diet and, 139–141

Oatmeal, 91
Oatmeal Waffles, 189–190
Obesity. *See* Weight gain
Oils, 121–123, 140
Olive oil, 122
Opiates, 32–33, 34, 36
Orange Sauce, Carrots in, 255–256
Ornish, Dean, 49, 121, 146
Osteoporosis, 37, 131, 136, 145, 153
Overeaters Anonymous, 46
Overeating, 69
Oyakata, Azumazeki, 56

Pancakes
Buckwheat Bananacakes, 187–188
Cornmeal Flapjacks, 188–189
Multigrain Pancakes, 186
Whole Wheat Pancakes, 185
Pasta, 60, 69
blood sugar and, 89, 90
Lasagne Roll-ups, 291–292
Pasta Salad, 219
product guide/resources, 312
protein from, 104
Simple Pasta Supper, 278–279
Pasta sauce, 312
Peaches
Fresh Peach Crisp, 294–295
Fresh Peach Freeze, 293
Ginger Peachy Bread Pudding, 301–302
Pears (Poached) with Butterscotch Sauce, 302–303

Peas
Curried Cauliflower with Peas, 254–255
Split Pea Soup, 246–247
Peppers
Portabello and Red Pepper Wraps, 233
Red Pepper Hummus, 197
Personality, taste and, 28
Phenylethylamine (PEA), 32
Physicians Committee for Responsible Medicine, 4, 158
Pima Indians, 16, 20, 21, 74
Pineapple Apricot Sauce, 204
Pita Pizzas, 235–236
Pizzas
Pita Pizzas, 235–236
Polenta Pizza, 276–277
Plum Sauce, 205
Polenta
basic recipe, 273–274
Breakfast Scramble, 183–184
Polenta Pizza, 276–277
Portion explosion, 69
Potato chips, 36
Potatoes, 68–69
blood sugar and, 89, 90, 91
Curried Potatoes and Chickpeas, 268–269
Double Potato Soup, 241
low-fat options, 60–61
Mashed Potatoes and Gravy, 252
Mashed Potatoes with Black Beans, 250
Potato Enchiladas, 281–282
Potatoes and Kale, 253
Potato Salad, 215–216
Skillet Scalloped Potatoes, 251
Prader-Willi syndrome, 13
Pregnancy, 148

Index

Index

changing one's taste for, 109
chocolate addiction, 31–34, 37
infants' attraction to, 148, 150
PROP tasters and, 27
taste for, 19, 30–31
weight gain and, 30–31

Tabouli, 216–217
Tamale Pie, 282–283
Tapioca
Apricot Tapioca, 300–301
Tapioca Pudding, 300
Taste, 3, 18
carbohydrate craving, 34–35
chocolate addiction, 31–34, 37
craved foods, 32
determining your taste type, 26
diet makeover and, 108–109, 110,
116–117
fatty foods, attraction to, 36–37
gender differences, 23–24
in infants, 24–25, 31, 148, 150
maintenance of, 108
nontasters, 25, 26, 28–29, 117
personality and, 28
premenstrual cravings, 37–40, 128
PROP tasters, 18–19, 25–28, 117
purpose of, 29–30
reeducating your taste buds, 108–109,
110
salty foods, attraction to, 37
sweets, attraction to, 19, 30–31
Teff, Breakfast, 181
Television watching, 155
Tempeh Salad Sandwich, 226
Texas Caviar, 210
Thai Wraps, 230–231
Theobromine, 32
Thermic effect of food (TEF). *See* After-
meal burn

Thorp, Edward, 17
Thyroid hormone, 75
Tofu
Missing Egg Sandwich, 226–227
Tofu, Lettuce, and Tomato Sandwich
(TLT), 230
Tofu Vegetable Hash, 290
Vicki's Tofu Mayo, 199
Tomatoes
Tofu, Lettuce, and Tomato Sandwich
(TLT), 230
Tortillas, 311
Treadmills, 101
Tryptophan, 35
Turkey, 35
Twin studies, 11–12
Type I and Type II muscle cells, 95–96

Uncoupling protein 2 (UCP2), 74
Unfamiliar foods, listing of, 173–177

Van Gogh, Vincent, 42
Vegetable dishes
Beets in Dill Sauce, 248
Broccoli with Sesame Salt, 256
Brussels Sprouts in Creamy Sauce,
258–259
Carrots in Orange Sauce, 255–256
Curried Cauliflower with Peas,
254–255
Green Beans with Garlic, 260–261
Mashed Potatoes and Gravy, 252
Mashed Potatoes with Black Beans, 250
Oven Roasted Vegetables, 256–257
Potatoes and Kale, 253
Ratatouille, 254
Sesame Kale, 249
Skillet Scalloped Potatoes, 251
Steamed Vegetables with Sesame Salt,
259–260

Index   349

❋ Conversion Chart ❋
Equivalent Imperial and Metric Measurements

American cooks use standard containers, the 8-ounce cup and a tablespoon that takes exactly 16 level fillings to fill that cup level. Measuring by cup makes it very difficult to give weight equivalents, as a cup of water will weigh considerably more than a cup of flour. The easiest way therefore to deal with cup measurements in recipes is to take the amount by volume rather than by weight. Thus the equation reads:

1 cup = 240 ml = 8 fl. oz.
½ cup = 120 ml = 4 fl. oz.

It is possible to buy a set of American cup measures in major stores around the world.

LIQUID MEASURES

Fluid Ounces	U.S.	Imperial	Milliliters
	1 tsp	1 tsp	5
¼	2 tsp	1 dessert-spoon	10
½	1 tbs	1 tbs	14
1	2 tbs	2 tbs	28
2	¼ cup	4 tbs	56
4	½ cup		120
5		¼ pint or 1 gill	140
6	¾ cup		170
8	1 cup		240
9			250, ¼ ltr
10	1¼ cups	½ pint	280
12	1½ cups		340
15		¾ pint	420
16	2 cups		450
18	2¼ cups		500, ½ ltr
20	2½ cups	1 pint	560
24	3 cups or 1½ pints		675
25		1¼ pints	700
27	3½ cups		750, ¾ ltr
30	3¾ cups	1½ pints	840
32	4 cups or 2 pints or 1 quart		900
35		1¾ pints	980
36	4½ cups		1000, 1 ltr
40	5 cups	2 pts or 1 qt	1120

SOLID MEASURES

U.S. and Imperial		Metric	
Ounces	Pounds	Grams	Kilos
1		28	
2		56	
3½		100	
4	¼	112	
5		140	
6		168	
8	½	225	
9		250	¼
12	¾	340	
16	1	450	

OVEN TEMPERATURE EQUIVALENTS

F	C	Gas Mark	Description
225	110	¼	Cool
250	130	½	
275	140	1	Very Slow
300	150	2	
325	170	3	Slow
350	180	4	Moderate
375	190	5	
400	200	6	Moderately Hot
425	220	7	Fairly Hot
450	230	8	Hot
475	240	9	Very Hot
500	250	10	Extremely Hot

Any broiling recipes can be used with the grill of the oven, but beware of high-temperature grills.

EQUIVALENTS FOR INGREDIENTS

all-purpose flour—plain flour
arugula—rocket
beet—beetroot
coarse salt—kitchen salt
cornstarch—cornflour
eggplant—aubergine
fava beans—broad beans
granulated sugar—caster sugar
lima beans—broad beans
scallion—spring onion
shortening—white fat
snow pea—mangetout
squash—courgettes or marrow
unbleached flour—strong, white flour
vanilla bean—vanilla pod
zest—rind
zucchini—courgettes or marrow
baking sheet—oven tray
plastic wrap—cling film

"Dr. Neal Barnard is one of the most responsible and authoritative voices in American medicine today."

ANDREW WEIL, M.D., AUTHOR OF *8 WEEKS TO OPTIMAL HEALTH* AND *SPONTANEOUS HEALING*

"No one knows more about healthy diets than Dr. Neal Barnard. I personally follow his recommendations faithfully."

HENRY J. HEIMLICH, M.D., THE HEIMLICH INSTITUTE AT DEACONESS HOSPITAL, CINCINNATI, OHIO

MAKE FOOD ONE OF YOUR PRIME TOOLS FOR MAINTAINING GOOD HEALTH WITH THESE OTHER BOOKS BY NEAL BARNARD, M.D.

Eat Right, Live Longer
0-517-88778-9. $13.00 paperback
(Canada: $17.95)

Food for Life
0-517-88201-9.
$14.00 paperback
(Canada: $21.00)

Foods That Fight Pain
0-609-80436-7. $14.00 paperback
(Canada: $19.50)

THREE
RIVERS
PRESS

Available wherever books are sold.